# BRITISH MINOR EXPEDITIONS.

## 1746 to 1814.

---

COMPILED IN THE INTELLIGENCE BRANCH OF THE QUARTERMASTER-GENERAL'S DEPARTMENT.

---

LONDON:

*Printed under the Superintendence of Her Majesty's Stationery Office, and sold by*

W. Clowes & Sons, Limited, 13, Charing Cross; Harrison & Sons, 59, Pall Mall;
W. H. Allen & Co., 13, Waterloo Place; W. Mitchell, 39, Charing Cross;
Longmans & Co., Paternoster Row; Trübner & Co., 57 & 59, Ludgate Hill;
Stanford, Charing Cross; and Kegan Paul, Trench, & Co., 1, Paternoster Square;
Also by Griffin & Co., The Hard, Portsea;
A. & C. Black, Edinburgh;
Alex. Thom & Co., Limited, Abbey Street, and E. Ponsonby, Grafton Street, Dublin.
1884.

*Price Two Shillings and Sixpence.*

LONDON:
PRINTED FOR HER MAJESTY'S STATIONERY OFFICE,
BY HARRISON AND SONS, ST. MARTIN'S LANE,
PRINTERS IN ORDINARY TO HER MAJESTY.

(Wt. 11738 650 10 | 84 717)

# CONTENTS.



*Sketch Maps to illustrate the various accounts.*

# PREFACE.

---

THIS volume contains a brief record of some of the expeditions which were despatched from the shores of England at various periods during the latter part of the eighteenth, and earlier portion of the nineteenth centuries.

These short accounts have been drawn up during intervals of more serious work by officers employed in the Intelligence Branch without any view to publication, and have no claim to originality, being for the most part compilations from well-known authorities: they merely serve to place before the reader in a handy shape the experience gained by former generations in this class of military operations.

INTELLIGENCE BRANCH,
*May*, 1884.

# BRITISH MINOR EXPEDITIONS.

## EXPEDITION TO LORIENT, 1746.

1746. Expedition to Lorient.

AFTER the capture of Cape Breton, in 1745, the British Government began preparations for an expedition against Quebec, the capital of Canada, then a French colony. With this view commissions were sent to the British Colonial Governors on the American continent to raise companies to join a force that was to be despatched from England. Eight thousand troops were raised in consequence of these directions; and a powerful squadron, with transports carrying six regiments, was got ready at Portsmouth. The sailing of the armament was, however, unaccountably delayed till late in the year; and it was then considered dangerous to risk ships on the exposed coast of North America. That these preparations might not be wholly useless to the nation, a new direction was given to the enterprise. It was thought that Lorient, the repository of all the stores and ships belonging to the French East India Company, might be surprised, and it was determined to make a descent on the coasts of Brittany, and thus facilitate the operations of Austria in the South of France, and of the allied armies in Flanders.

The force sailed from Plymouth on the 14th September, 1746. It consisted of 9 men-of-war, 2 sloops, a fire-ship, bomb-ketch, some small armed vessels, and 33 transports, having on board a battalion of the Royals, and Bragg's, Harrison's, Frampton's, Richbell's, and Lord John Murray's regiments. Admiral Lestock commanded the fleet, and General Sinclair the land forces. Two men-of-war and a sloop, previously despatched from Plymouth to reconnoitre the coast and select a landing-place, joined the fleet off the Island of Groix on the 18th. On the 19th the fleet anchored in Quimperlé Bay, a distance of about three leagues from Port Louis, and rather more from Lorient.

The expedition took the French entirely by surprise, but the alarm was at once given by firing guns from Port Louis, and by night-signals, bonfires, &c., in order to raise the country against the British. At daybreak on the morning of 20th September the Admiral made the signal for the boats of the fleet to assist in the disembarkation, when a force consisting of Militia, armed peasants, and three or four troops of Dragoons from Port Louis, numbering in all about 3,000, was seen on the coast. In order to cover the disembarkation, three 40-gun ships, two sloops, and some small vessels were ordered close in shore. These vessels kept up a continuous fire on the enemy for some hours, but not much execution

1746. Expedition to Lorient.

was done, as the main body retreated behind some rising ground on the first broadside.

In the meantime, the boats of the squadron were pulling backwards and forwards, watching for a favourable opportunity to land; and the troops were the more eager to get ashore as the French had brought down two 6-pounders to a point about half-a-mile distant, and opened fire with grape. Before this the boats had made a feint towards the shore, and the French immediately drew up to dispute the landing. As soon as the General saw them in motion he made a signal for the boats to pull to leeward, and by this manœuvre succeeded in landing 600 men about half-a-mile nearer Port Louis before the French could get into position to oppose them. The landing once effected the enemy fled in all directions. The disembarkation was quickly completed, and in the afternoon the force, under the guidance of some peasants, marched to the village of Guidel. The same evening the marines were landed for the protection of the guns, and to keep open communication with the fleet. Including these, the whole effective force amounted to 4,500 men.

On the 21st September an advance was made to Plœmeur, six miles from the place of landing, and about four from Lorient. The closeness of the country enabled the peasants to annoy the British force from the thickets and hedgerows along the road and several men were killed and wounded. At one point a body of Militia and armed peasants had taken up their position and opened fire upon Frampton's and Richbell's regiments, which, never having been in action, gave way in disorder with the exception of the grenadier company of Frampton's, which stood its ground. The troops were however promptly rallied, whereupon the French fled.

On the 22nd the British force advanced to within two miles of Lorient, and the place was summoned to surrender. The governor, at his own request, was allowed till 7 o'clock the next morning to consider his answer.

At that hour a flag of truce arrived at the outposts with a proposition to the effect that the garrison of Lorient should march out with all the honours of war, and be then permitted to disband.

General Sinclair, however, having had the most confident assurances from his engineer officers, after a reconnaissance of the town, that in twenty-four hours they could with the available artillery effect a breach, or lay the town in ashes, rejected these conditions, and insisted on a surrender at discretion. A second flag of truce asking for less severe terms came out three hours after; but the General, relying on the judgment of his engineer officers, would listen to no terms but unconditional surrender, and negotiations were forthwith at an end.

Preparations were now made for throwing up earthworks to bombard the town, and two 12-pounders and a 10-inch mortar were dragged up from from the beach by the marines and sailors. On the 24th October, the British force encamped within cannon shot of the place, and the same night the construction of a battery was commenced under the protection of a covering party of 700 men. Before daybreak the work was completed, and the two guns

## EXPEDITION TO LORIENT 1746.

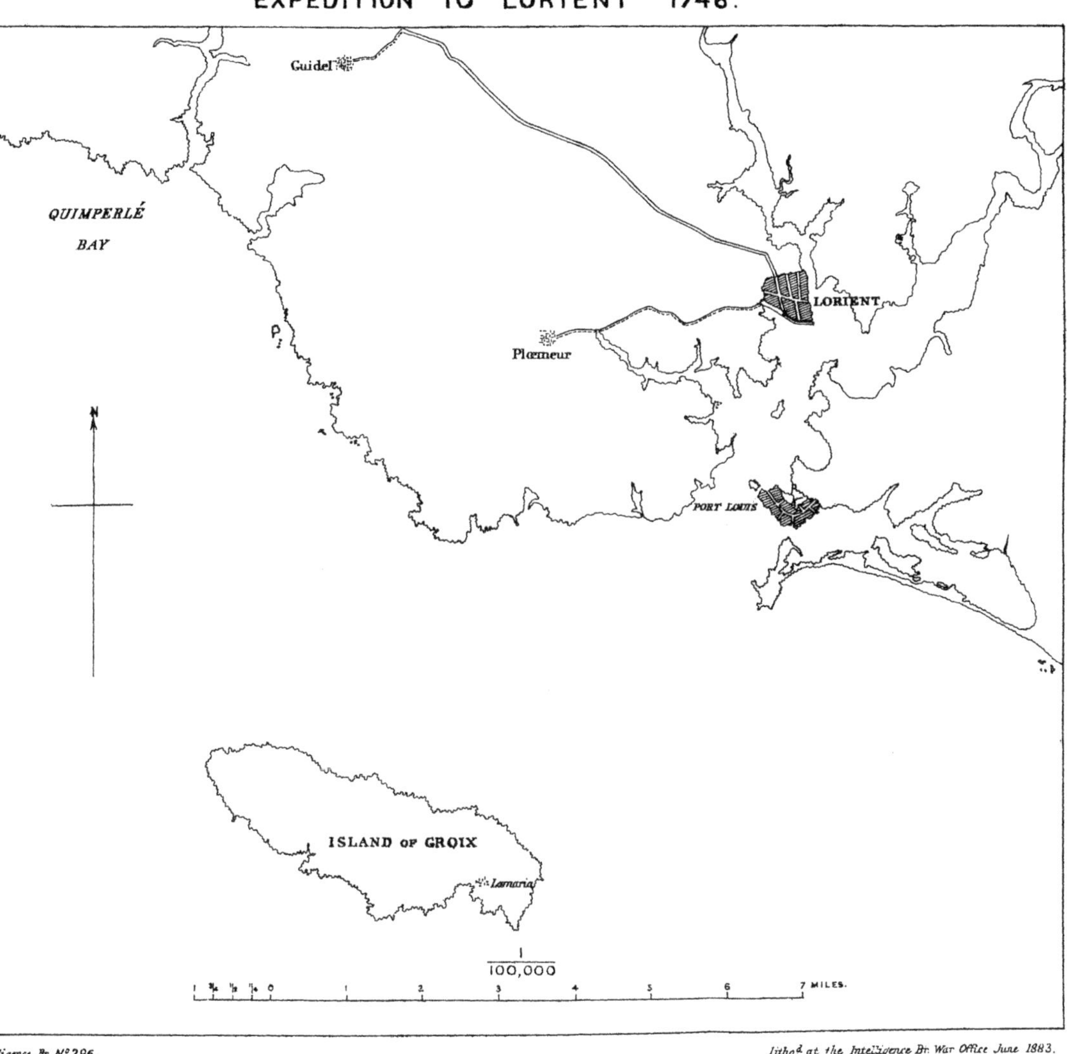


llligence Br. No. 296.

Lithod. at the Intelligence Br. War Office June 1883.

and mortar were mounted. The battery was in a favourable position on the side of a hill about 300 yards from the town; but it was soon evident that the French had been no less industrious on their part, for in the morning, when the bombardment began, the fire was returned from seven pieces of artillery, and a mortar of much heavier metal than that of the besiegers.

1746. Expedition to Lorient.

With the British, matters went from bad to worse. The ammunition came up very slowly on account of the badness of the road and the want of horses. Two more 12-pounders were mounted on the 25th October, but it was impossible to keep up a continuous fire for want of ammunition. The troops having been under arms ever since the landing, suffered greatly from overwork, which was aggravated by the inclemency of the weather, so that numbers of men became daily unfit for duty. And to add to these complications, Admiral Lestock represented that the fleet lay in a most dangerous position, entirely exposed to the west and south-west, and in the event of a gale from either of those quarters there was great risk of the ships going ashore.

The French, on the other hand, had more guns, and constant supplies of men, ammunition, and all necessaries from Port Louis and other neighbouring garrisons, and maintained an incessant fire.

In these circumstances a council of war was assembled, and it was unanimously agreed that the troops should be immediately re-embarked. Accordingly, on the evening of the 26th September, the four guns and mortar were spiked, and the force retreated to the seashore, whence the embarkation was effected without molestation on the 27th and 28th.

Little damage had been done to Lorient; a few houses were set on fire by red-hot shot, but as the town was for the most part built of stone, the flames were speedily extinguished.

The British loss was some 10 or 12 men killed, and about 40 prisoners.

On the 1st October the fleet sailed from Quimperlé Bay, and heavy weather being experienced, some of the vessels were nearly wrecked on the rocks of the Isle of Groix. Owing to this the fleet got somewhat dispersed, and no rendezvous having been appointed, on the French coast, some of the transports sailed for Falmouth. The whole expedition was on the point of returning to England, as there was nobody in the fleet capable of piloting it into any other French port, when a tender joined it with pilots from Guernsey and Jersey. A council of war was then held by the Admiral, and it was resolved to proceed with the fleet to Quiberon. The new rendezvous was given to all the ships that could be spoken with.

On the 2nd of October the fleet cast anchor at the entrance of Quiberon Bay. On the 3rd it moved further in, and anchored within a league of the peninsula of Quiberon, where the troops landed the next day without opposition. Upon this peninsula is the town of Lomaria (now called Quiberon) and twenty-five villages, which were all deserted by the inhabitants, who carried off most of their property. Eighteen pieces of heavy cannon were found here;

1746. Expedition to Lorient.

seven of these were at once mounted to command the isthmus, on which strong intrenchments were thrown up.

The British remained stationary in this position for several days, but as there was no news of the missing transports which had parted from the fleet with 900 men on board, nor any hope of reinforcements, it was thought expedient to re-embark the troops, as the force only amounted to 2,500 men, and it would have been rash to have attempted to penetrate further into the country with so small a force. The troops therefore re-embarked on the 11th October, having previously spiked the guns; thirty or forty brigs, sloops, and barges were also destroyed.

While the troops were at Quiberon, Admiral Lestock despatched two or three men-of-war to cannonade the islands of Haedik and Houat. A body of sailors was landed at the same time, and the garrisons, consisting of some sixty men, surrendered. About 500 head of cattle were brought off and distributed among the ships of the fleet; the forts were then destroyed, after which exploit the fleet returned to England.

The only effect of this feeble and fruitless expedition was to exasperate the French people. To land a handful of troops without horses, tents, or artillery, from a fleet lying in an open and exposed roadstead like that of Quimperlé, at a time of the year when heavy weather was to be expected, and thus render a re-embarkation precarious, was a useless and dangerous proceeding not likely to distress the French, but almost certain to prove disastrous to the expedition.

---

## FIRST EXPEDITION TO ST. MALO AND CHERBOURG, 1758.

1758. First Expedition to St. Malo and Cherbourg.

In the spring of 1758 Mr. Pitt directed that an expedition should be organised to act upon the coast of France. The Government had two objects in view: the one, to destroy the enemy's principal naval arsenals, and burn, sink, or capture his shipping; the other, to create a diversion in favour of the King of Prussia, by compelling the French to turn their attention to the protection of their coast.

A force, composed as follows, amounting to about 13,000 men, was collected in the Isle of Wight:—

Three battalions of Guards, and the 5th, 20th, 23rd, 24th, 25th, 30th, 33rd, 34th, 36th, 67th, and 68th Regiments. Nine troops of Light Cavalry, and 400 artillerymen, with about 60 guns.

The command of the land forces was given to the Duke of Marlborough, whose staff consisted of Lieut.-Generals Lord George Sackville and Lord Granby; Major-Generals Waldegrave, Mostyn, Drury, Boscawen, and Elliot; Brigadier Elliot commanding the Light Horse; Lieut.-Colonel Hotham Adjutant-General, and Captain Watson Quartermaster-General. Admiral Lord Anson and Vice-Admiral Sir Edward Hawke commanded the covering squadron, the transports and vessels of light draught being placed under the orders of Commodore Howe.

On the 1st of June the convoy sailed from St. Helens, Lord Anson and the fleet having preceded them; but the weather becoming tempestuous, it was not until the 5th that anchor was cast in the Bay of Cancale. The transports, having the Guards on board, were ordered to stand in near the shore, protected by three of the frigates. The troops were pulled towards the shore in flat-bottomed boats, when a battery opened fire upon them at short range; but the well-directed fire from the frigates soon silenced the enemy's guns, and the troops soon after landed without opposition, and seized the village of Cancale.

**1758. First Expedition to St. Malo and Cherbourg.**

The remainder of the troops were now landed, and on the following morning the force, with the exception of the 3rd Brigade, which remained at Cancale, marched towards St. Malo, and encamped within a mile of the town. The Light Horse were sent on to reconnoitre, and advanced close to the walls of the town when they were fired upon and at once retired without loss.

At night Brigadier Elliot, with 200 cavalry and a similar number of infantry, *en croupe*, carrying hand grenades, made his way to the harbour of St. Servan, where he found a 50-gun ship, two 36-gun frigates, upwards of twenty privateers, and seventy or eighty merchant ships: these and the magazines of pitch, tar, and other naval stores were set on fire by the troops. The garrison of St. Malo made no attempt at rescue, and the whole were burnt.

On the 8th June preparations were made for besieging the town, and the 2nd Brigade, with two guns, under Major-General Waldegrave, moved into St. Servan to dislodge a small force of the enemy which had taken refuge in a fort on the point opposite St. Malo. Meanwhile, the Duke of Marlborough reconnoitred the town, and decided that it was too strong to be carried by a *coup de main*, and would take at least a month to reduce by siege. While debating as to the course to pursue, news arrived of the concentration and advance of the enemy to cut off his retreat. Orders were at once issued to re-embark, and by the 12th all were on board.

The fleet, after remaining at anchor until the 16th, moved along the coast to the point of Roteneuf; but the next day, owing to stress of weather, returned to Cancale Bay. In the meantime, the military and naval commanders having reconnoitred Granville, decided not to risk an attack.

On the 21st the convoy again stood out, and was driven by a westerly gale to within sight of the Isle of Wight. When the gale abated, they stood back for the French coast, which was sighted about 4 P.M. on the 25th. On the 26th, the fleet was off the mouth of the Seine and, sailing along close in shore, arrived off Cherbourg on the 29th, and anchored about two miles from the town. Six batteries opened fire, but without effect, and considerable numbers of troops were seen on the beach.

The Duke of Marlborough decided to land, while the ships bombarded the forts of Querqueville, Homet, and Galet. The 1st battalion of Guards, supported by the Grenadiers, were to form

1758. First Expedition to St. Malo and Cherbourg.

the first or storming party; the right companies, under Colonel Carey, were to attack Fort Querqueville; the left, under Lord F. Cavendish, were to assault Fort Homet with the battery between it and Fort Galet; the centre companies, under Colonel Pierson, were ordered to take the battery between Homet and Querqueville, and if possible to maintain their position on shore till the morning, when the remainder of the troops were to be landed.

It was supposed that the town was garrisoned by a regiment which from its uniform was taken to belong to the Irish Brigade. The batteries were manned by Militia, and in rear of them a number of men, in strength equal to about two battalions, were seen under arms. Small parties of cavalry were also observed at different points.

A portion of the Guards had already entered the boats, but the wind springing up suddenly, the men were ordered back to their transports. The gale continued to increase, and blew directly into the bay; the transports in working out ran foul of each other, and with difficulty escaped shipwreck. It was therefore determined to return to England, and on the 1st of July the fleet arrived at St. Helens.

## SECOND EXPEDITION TO CHERBOURG, 1758.

Second Expedition to Cherbourg.

THE destruction of the shipping and stores at Cherbourg was considered of sufficient importance to warrant a further attempt. Consequently, after a rest of about a fortnight, the troops were again ordered to embark, under the command of Lieut.-General Bligh, the fleet as before being under the orders of Commodore Howe.

On this occasion the instructions received from the King pointed specifically to the destruction of Cherbourg.

"Our will and pleasure is, that you do exert your utmost endeavours to land, if it should be found practicable, with the troops under your command, at or near Cherbourg, on the coast of Normandy, and to attack the batteries, forts, and town of Cherbourg; and in case, by the blessing of God upon our arms, the said place shall be carried, and that our troops shall be able to maintain themselves there a competent time, for demolishing and destroying the port and basin, together with all the ships, naval stores and works, batteries, fortifications, arsenals, and magazines thereunto belonging, you are to use all possible means effectually to demolish and destroy the same," &c.; and should the attempt on Cherbourg not succeed, "to carry a warm alarm along the coast of France, from the easternmost point of Normandy, as far westward as Morlaix inclusive."

Thanks to the energy of Commodore Howe, the ships were revictualled, and everything was in readiness for the reception of the troops on the 23rd July.

The fleet sailed from Spithead on the 30th July, but owing to light and variable winds, could not get clear of the land until the 1st of August, and did not reach the French coast near Cherbourg until the afternoon of the 6th. 1758. Second Expedition to Cherbourg.

At this time Cherbourg was weakly defended, and the works begun by Louis XIV had not been carried on by his successor. The town itself was open, and the coast defences were confined to some scattered forts and batteries along the shore from Tourlaville to Querqueville. Along the eastern shore there was an intrenchment, and a redoubt occupying the site near which the Fort des Flamands now stands. To the west of the town there was a fort called Langlet, and at intervals of 600 to 700 yards further on stood—

1. The Fort du Galet, a work with two faces to the sea, two flanks to the beach, and a hornwork closing the gorge.
2. Homet, a square redoubt, near the point of the same name.
3. Equerdreville, a battery *en barbette* facing the sea, with towers commanding beach and land.

A little above this work a line of intrenchments with several batteries stretched along the coast, from St. Anne to Querqueville, at which point there was a work of similar shape to that at Equerdreville.

The French had available for the defence of these works about 9,000 men, under the *Maréchal de camp* le Comte de Raimont.

| | |
|---|---|
| Lord Clare's Regiment of Irish .. .. | 700 |
| Le Comte de Lorrain's Regiment .. .. | 600 |
| Le Comte de Horon's Regiment .. .. | 1,300 |
| Gardes des Côtes .. .. .. .. | 6,000 |
| Dragons de Languedoc .. .. .. | 600 |
| | 9,200 |

At daybreak on the 7th of August the fleet made a demonstration against the forts. The transports moved further west, and preparations were made for landing. Some bomb vessels and frigates were ordered close in shore to sweep the beach with their fire, and the flat-bottomed boats with the troops were towed towards Erville, where from the rocky nature of the coast the French had not supposed that a landing would be made. The enemy had, however, seen the direction taken by the transports in the morning, and had sent troops to the menaced point, 3,000 of whom had taken up a position behind the sandhills.

The 1st division of Guards and Grenadiers, 500 strong, landed under cover of a heavy fire from the ships, and advanced against the French without returning their fire. The enemy retired, leaving a pair of colours and two guns in the hands of the English, who lost about twenty men killed and wounded. The coast being now clear, the remainder of the troops disembarked.

General Bligh encamped for the night at Erville, detaching one regiment to Nacqueville. The position was cramped, and had the enemy attacked in force, the struggle would have been severe, but no attempt was made to prevent the further landing of the troops, which was continued on the following morning.

1758. Second Expedition to Cherbourg.

The advance on Cherbourg was made in two columns, preceded by an advanced guard of cavalry and grenadiers with two guns, which followed the road running along the low ground to Querqueville. The main body took the same road, and the second column crossed over the hills from Nacqueville.

The two columns united at Fort Querqueville, which had been abandoned, and detaching a party of cavalry to Hainville, marched in a single column along the road, passing behind Le Galet to Cherbourg. The advance was unopposed, and the scouting party from Hainville found the country deserted by the enemy. It was midnight before the tail of the column reached Cherbourg, where the officers billeted themselves on the inhabitants, leaving the men to shift for themselves.

Next day the town and neighbourhood were examined, when to the surprise of all it was found that the enemy had retired. The work of destruction was then carried out, the demolition of the forts being left to the Engineers, and that of the basin to the fleet and Artillery.

Twenty-seven ships were burnt in the harbour, and 173 pieces of iron ordnance and some mortars were destroyed. Twenty-four brass guns, and the colours taken on landing, were sent home.

While the forts and basin were being destroyed, lines were made to cover an embarkation if necessary, and the Cavalry was sent out in various directions. They soon ascertained that the French were concentrating at Valognes, a town situated only some twelve miles south-east of Cherbourg. Skirmishes were of daily occurrence, and several deserters from the Irish regiment came into the English camp and gave information to the effect that the force at Valognes was under the Duke of Luxembourg, and, in addition to the Cherbourg garrison, was composed of—

| | |
|---|---|
| The *Régiment* Guienne .. .. .. | 700 |
| „ Limousin .. .. .. | 700 |
| „ Poictou .. .. .. | 1,400 |
| Cavalry .. .. .. .. .. | 1,500 |
| | 4,300 |

making in all a total force of 13,500.

The destruction of the basin was completed on the 15th of August, and the forts and magazines having been destroyed some days before, General Bligh determined to re-embark, which he was allowed to do unmolested.

On the 18th August the fleet put to sea with the intention of attacking either Granville or Morlaix, but advices were received from England that a large force had assembled near Brest ready to oppose any descent on the latter part of the coast. It was therefore decided to proceed to the Bay of St. Lunaire, four miles west of St. Malo, and on the 7th September the troops landed without opposition.

The Grenadiers burnt the shipping in the harbour of St. Briac, and a council of war was held on the subject of attacking St. Malo. The weather, however, being very unsettled, the Commodore stated

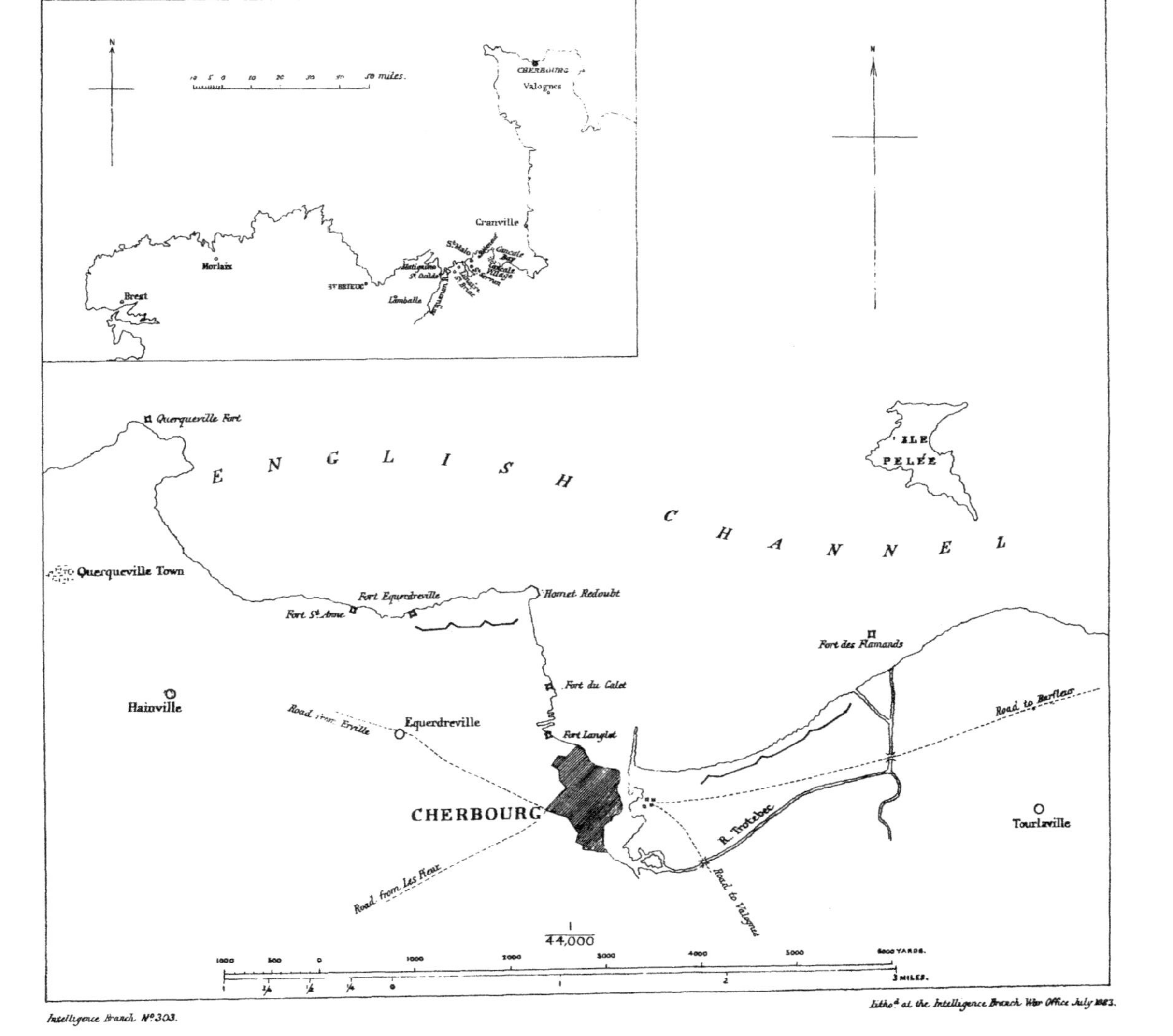
Querqueville Fort
ENGLISH CHANNEL
ILE PELÉE
Querqueville Town
Fort Equerdreville
Fort St. Anne
Homet Redoubt
Fort des Flamands
Fort du Calet
Fort Langlet
Hainville
Equerdreville
Road from Erville
CHERBOURG
Road to Barfleur
R. Trotebec
Tourlaville
Road from Les Pieux
Road to Valognes
1/44,000
YARDS
MILES
Brest
Morlaix
Granville
St. Malo
Lamballe
Valognes
CHERBOURG
miles
Litho.d at the Intelligence Branch War Office July 1883.
Intelligence Branch No. 303.

that it was impossible for him to assist the troops by bombarding the town, as he would thereby run the risk of losing the whole of the fleet, and that in consequence of the westerly gales it was impossible to remain any longer in the Bay of St. Lunaire. To enable the troops to re-embark, he therefore proposed to move round to the more sheltered Bay of St. Cas, which the troops could reach by land.

1758. Second Expedition to Cherbourg.

The force moved early on the morning of the 9th September towards Matignon, but on reaching the river Arguenon in the afternoon, it was found impassable owing to the tide. On the following day at low water the Grenadiers of the line crossed the ford opposite St. Guildo, and the Grenadiers of the Guards tried to pass lower down opposite the wood of Val, which was occupied by the French, who opened fire as soon as the Guards entered the river. They, however, effected a crossing, and driving the enemy before them, took the village, but not without loss. The main body of the Brigade of Guards found a better passage lower down, and forded the river without difficulty; the other Brigades followed the same route, and by nightfall all were across.

On the 11th the English marched to Matignon, three miles from St. Cas, where the fleet was at anchor. General Bligh was now informed that the French from St. Brieuc, to the number of 14 battalions, 4 squadrons, and 12 guns, were assembling between Lamballe and Matignon ; and that another body of 3,000 men, under D'Aubigny and La Chartre, was advancing from St. Malo. On receipt of this intelligence General Bligh, with the concurrence of most of the General officers, gave orders for the re-embarkation on the following day.

At daybreak on the 12th September, the English moved off to St. Cas, with a rear-guard of four companies of the 1st Guards, and the Grenadier companies of the line. The enemy interfered but little during the march, and the troops began to embark at 9 o'clock ; by 11 o'clock the three Brigades of the line, half the Guards, and the wounded were on board. The enemy now appeared in force on the heights, and at once opened fire on the British rear-guard under General Drury, amounting to about 1,500 men. The commander drew his men up at once behind a natural breastwork, and being supported by the fire of the fleet, at first held his own. Tempted, however, by the failure of a French attack, the English left their cover, and before they could form up, were overwhelmed by superior numbers and driven back in confusion. A rush was now made for the boats, but as many of these had been destroyed by the enemy's artillery, only a portion of the rear-guard was able to get away. Many were cut down on the beach, several were drowned, and General Drury and 400 men were taken prisoners—the total loss being over 700 men.

With the force at his disposal, it was considered that the English commander would have done better to have risked an action, rather than attempt an embarkation with an enemy in force and so near.

1761. Siege and capture of the Island of Belleisle.

## SIEGE AND CAPTURE OF THE ISLAND OF BELLEISLE.

EARLY in the year 1761 the English Government determined to send an expedition to seize Belleisle. This island lies off the coast of Brittany abreast of Quiberon Bay, and from its size and position was well suited as a point from which to threaten the mainland.

The transports were collected at Spithead, about the beginning of March, where they were joined by the "Sandwich," ninety guns; the "Valiant," "Torbay," "Téméraire," "Dragon," and "Swiftsure," each of seventy-four guns; the "Prince of Orange," seventy guns; the "Achilles," sixty guns; and several frigates, the whole convoy being under the command of Commodore Keppel.

The land forces, amounting to about 10,000 men, were under the orders of General Hodgson and Brigadiers Howe, Carlton, and Desaguillier, and were composed as follows:—

The 16th Light Dragoons; the 9th, 19th, 21st, 30th, 67th, 69th, 76th, 85th, 90th, 97th, and 98th Regiments.

The garrison of Belleisle was commanded by the Chevalier de St. Croix, and was composed of the Régiment de Nice (2 battalions), the Régiment de Bigorre, the Bataillon de Dinan, which, with the Artillery, Engineers, and Coast Guard, formed a force of 150 officers and 2,500 men.

Every advantage had been taken of the rocky and precipitous nature of the coast to place the island in a perfect state of defence, and the works of Le Palais, the citadel, had been but lately strengthened. The three harbours of Palais, Sauzon, and Goulfard were of little use to the fleet, being all more or less exposed, shallow, or dangerous at the entrance.

The expedition started on the 29th of March, and arrived off Belleisle about noon on the 7th April. Preparations were immediately commenced for landing the troops by getting the flat-bottomed boats in readiness; meanwhile the Commodore and General Hodgson proceeded to the north of the island, to inspect the defences.

On the 8th April, at an early hour, signal was made for a part of the troops to land, and the "Achilles," "Prince of Orange," and "Dragon," with the "Fire Drake" and "Furnace" bombs, and the "Escorte" frigate, were sent round the point of Lomaria, at the south of the island, to attack a fort and other works situated there. The guns of the "Achilles" soon silenced the fire of the fort and shore batteries, when the boats containing the Grenadiers and other troops pushed off; but on landing it was found impossible to reach the works, owing to the steepness of the ground, which had at that point been scarped. After several unsuccessful attempts to close with the enemy, in one of which sixty Grenadiers of Erskine's succeeded in forcing their way to the top of the cliff, where they were at once overpowered, the troops retired, having lost in killed, wounded, and prisoners, about 400 men.

With a view to creating a diversion. Sir Thomas Stanhope, with

# SIEGE & CAPTURE OF BELLEISLE 1761.

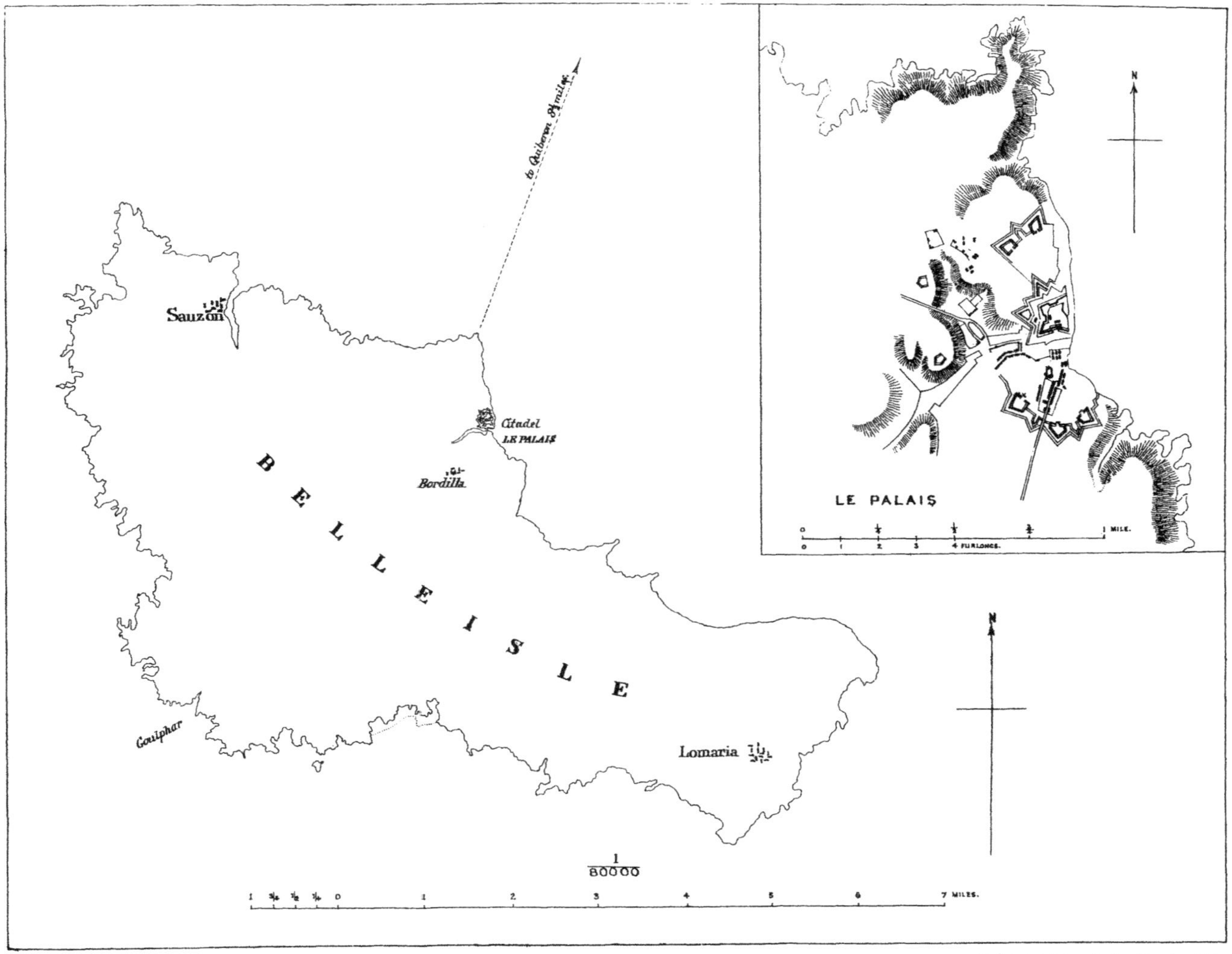

to Quiberon 8¾ miles.

Sauzon

Citadel

LE PALAIS

Bordilla

BELLEISLE

Goulphar

Lomaria

N

1/80000

1 ¾ ½ ¼ 0 1 2 3 4 5 6 7 MILES.

LE PALAIS

N

0 ¼ ½ ¾ 1 MILE.

0 1 2 3 4 FURLONGS.

Intelligence Br. Nº 297.

Lithoᵈ at the Intelligence Br. War Office June 1883.

four ships-of-war, the battalions of Grey and Stuart, and 500 marines were sent to Sauzon; but the troops were prevented from landing by the weather. Nothing was attempted for some days after, owing to the gale which commenced that evening, scattering the fleet and causing the destruction of a large number of flat-bottomed boats.

1761. Siege and capture of the Island of Belleisle.

On the 22nd April, at 5 P.M., all the forts and works were simultaneously attacked by the ships, and feints were made at landing on several points. While the enemy's attention was thus distracted, Brigadier Lambert with Beauclerk's regiment, landed at a part of the coast which the enemy considered unassailable, and had consequently neglected. A party led by Captain Patterson scrambled up the rocks, and forming on the top, repulsed the attack of some 300 men, who were promptly sent against them. They were soon after joined by the remainder of the regiment, and a detachment of marines, and moved forward, capturing three brass field pieces and some wounded. The whole of the troops now landed, and drove the enemy into Le Palais, where they were joined by the local Militia, raising the strength of the garrison to about 4,000 men.

The English forces pushed forward on the 23rd April, and on the following day the detachment of Light Horse was sent to Sauzon, a body of infantry was placed in the village of Bordilla, and the remainder of the army invested the place. The stormy weather, which continued for several days, prevented the landing of artillery and siege material, and was utilised by M. de St. Croix in adding to the already strong defences of the town.

About the end of the month some mortars were brought up and opened fire upon the town, to which the enemy replied from their well-armed batteries, and on the 3rd of May attacked the trenches with great determination, driving back the left piquets. Notwithstanding the efforts of General Crauford, who commanded at that point, the works were destroyed. A great number of men were killed, and the General and his aide-de-camp were taken prisoners. This attack on the left was not followed up, and, satisfied with their success, the French retired. In less than twenty-four hours the works were repaired and redoubts were constructed to guard against a second surprise.

From this time the siege was actively carried on, but the Engineers having pointed out that the lines could not be carried nearer to the place in consequence of the newly-constructed redoubts, it was settled that they should be assaulted on the 13th May. The attack commenced at daybreak by a heavy artillery fire directed against the work on the French right. After which a detachment of marines and part of Loudon's Regiment dashed over the parapet and seized the work. The other redoubts were carried successively by the same storming party reinforced by Colvill's regiment, and the enemy, after a gallant resistance, and severe loss, were driven into the citadel. The storming party penetrated into the town and released some of the prisoners.

The defence was now confined to the citadel, which was perfectly isolated, and it was evident that the place must surrender

1761. Siege and capture of the Island of Belleisle.

as soon as the provisions were consumed; still M. de St. Croix held out, and kept up a heavy fire upon the besiegers.

The siege works were carried nearer and nearer, and an incessant fire from mortars and artillery was kept up from the 13th until the 25th of May, when the enemy's fire slackened. By the end of the month a breach was made which was declared practicable on the 7th June, and the place was summoned prior to the final assault.

For two months the Governor had held out against great odds, and with the conviction that he could obtain no support; having, therefore, done all that honour demanded, he wisely determined to prevent unnecessary bloodshed by capitulation. A few days after, in accordance with the articles agreed to by the respective commanders, the garrison was carried to the coast of France.

Belleisle remained in the hands of the English until the peace of Paris in 1763.

---

## APPENDIX I.

*Capitulation for the Citadel of Belleisle, made June 7th,* 1761.

Capitulation of the Citadel of Belleisle.

Preliminary Article.—The Chevalier de St. Croix, Brigadier in the King's army, and Commandant of the citadel of Belleisle, proposes that the place shall surrender on the 12th of June, in case no succours arrive before that time; and that in the meanwhile no works shall be carried on on either side, nor any act of hostility, nor any communication between the English besieging and the French besieged.—Refused.

I. The entire garrison shall march through the breach with the honours of war, drums beating, colours flying, lighted matches, and three pieces of cannon, with twelve rounds each. Each soldier shall have fifteen rounds in his cartouch-box. All the officers, serjeants, soldiers, and inhabitants are to carry off their baggage; the women to go with their husbands.—Granted, in favour of the gallant defence which the citadel has made under the orders of the Chevalier de St. Croix.

II. Two covered wagons shall be provided, and the effects which they carry shall be deposited in two covered boats, which are not to be visited.—The covered wagons are refused; but care shall be taken to transport all the baggage to the Continent by the shortest way.

III. Vessels shall be furnished for carrying the French troops by the shortest way into the nearest ports of France by the first fair wind.—Granted.

IV. The French troops that are to embark are to be victualled in the same proportion with the troops of His Britannic Majesty; and the same proportion of tonnage is to be allowed to the officers and soldiers which the English troops have.—Granted.

V. When the troops shall be embarked a vessel is to be furnished for the Chevalier de St. Croix, Brigadier in the King's army; to M. de la Ville, the King's Lieutenant; to M. de la Garique, Colonel of foot, with brevet of Commandant in the absence of the Chevalier de St. Croix; and to the field officers, including those of the artillery and engineers; as also for the three pieces of cannon, as well as for the

1761. Siege and capture of the Island of Belleisle.

Capitulation of the citadel of Belleisle.

soldiers of the Cour Royale, to be transported to Nantes, with their wives, servants, and the baggage which they have in the citadel, which is not to be visited. They are to be victualled in the same proportion with the English officers of the same rank.—Care shall be taken that all those who are mentioned in this article shall be transported, without loss of time, to Nantes, with their baggage and effects, as well as the three pieces of cannon granted by the first Article.

VI. After the expiration of the term mentioned in the first Article, a gate of the citadel shall be delivered up to the troops of His Britannic Majesty, at which there shall be kept a French guard of equal number, until the King's troops shall march out to embark. Those guards shall be ordered to permit no English soldier to enter, and no French soldier to go out.—A gate shall be delivered to the troops of His Britannic Majesty the moment the capitulation is signed, and an equal number of French troops shall occupy the same gate.

VII. A vessel shall be furnished to the Commissaries of War, and to the treasurer, in which they may carry their baggage, with their secretaries, clerks, and servants, without being molested or visited. They shall be conducted, as well as the other troops, to the nearest port of France.—Granted.

VIII. M. de Taille, Captain-General of the Garde Côte, Lamp-Major, two lieutenants of cannoniers of the Garde Côte, and ninety bombardiers, cannoniers, serjeants, and fusiliers Gardes Côtes of Belleisle, paid by the King, shall have it in their choice to remain in the island, as well as all the other inhabitants, without being molested, either as to their persons or goods. And if they have a mind to sell their goods, furniture, boats, nets, and in general any effects which belong to them, within six months, and to pass over to the Continent, they shall not be hindered; but, on the contrary, they shall have proper assistance and the necessary passports.—They shall remain in the island under protection of the King of Great Britain, as the other inhabitants; or shall be transported to the Continent, if they please, with the garrison.

IX. M. Sarignon, Clerk of the Treasury of the French troops, the armourer, the Bourgeois cannoniers, the storekeepers, and all the workmen belonging to the engineers, may remain at Belleisle with their families, or go to the Continent with the same privileges as above mentioned.—Granted. To remain in the island upon the same footing with the other inhabitants, or to be transported with the garrison to the Continent, as they shall think proper.

X. The Roman Catholic religion shall be exercised in the island with the same freedom as under a French government. The churches shall be preserved, and the rectors and other priests continued, and in case of death they shall be replaced by the Bishop of Vannes. They shall be maintained in their functions, privileges, immunities and revenues.—All the inhabitants, without distinction, shall enjoy the full exercise of their religion. The other part of this Article must necessarily depend on the pleasure of His Britannic Majesty.

XI. The officers and soldiers who are in the hospitals of the town and citadel shall be treated in the same manner as the garrison, and, after their recovery, they shall be furnished with vessels to carry them to France. In the meanwhile they shall be supplied with subsistence and remedies till their departure, according to the state which the comptroller and surgeons shall give in.—Granted.

XII. After the term mentioned in the Preliminary Article has expired orders shall be given that the commissaries of artillery,

1761. Siege and capture of the Island of Belleisle.

Capitulation of the citadel of Belleisle.

engineers, and provisions shall make an inventory of what shall be found in the King's magazines, out of which bread, wine, and meat shall be furnished to subsist the French troops to the moment of their departure. They shall be furnished with necessary subsistence till their departure on the same footing with the troops of His Britannic Majesty.

XIII. Major-General Crauford, as well as all the English officers and soldiers who have been made prisoners since the 8th of April, 1761, inclusive, shall be set at liberty after the signing of the capitulation, and shall be disengaged from their parole. The French officers of different ranks, volunteers, serjeants, and soldiers, who have been made prisoners since the 9th of April, shall also be set at liberty. The English officers and soldiers, prisoners of war in the citadel, are to be free the moment the capitulation is signed. The French officers and soldiers who are prisoners of war shall be exchanged according to the cartel of Sluys.

All the above articles shall be executed faithfully on both sides, and such as may be doubtful shall be fairly interpreted.—Granted.

After the signature hostages shall be sent on both sides for the security of the articles of the capitulation.—Granted.

All the archives, registers, public papers, and writings, which have any relation to the government of this island, shall be faithfully given up to His Britannic Majesty's commissary. Two days shall be allowed for the evacuation of the citadel, and the transports necessary for the embarkation shall be ready to receive the garrison and their effects. A French officer shall be ordered to deliver up all the warlike stores and provisions, and, in general, everything which belongs to His Most Christian Majesty, to an English commissary appointed for that purpose; and an officer shall be appointed to show us all the mines and souterrains of the place.

S. Hodgson.
Le Chevalier de St. Croix.
A. Keppel.

---

## EXPEDITION TO QUIBERON BAY, 1795.

1795. Expedition to Quiberon Bay.

In the year 1794, the English Government, yielding to the representations of the Royalist *emigrés* in England, determined on sending an expedition to the south coast of Brittany. In the month of April of the same year, a sum of money was voted in Parliament for the expenses of the expedition, and steps were taken to raise certain regiments of the Royalist *emigrés* in England and Germany, and also to obtain recruits from the French sailors then prisoners of war in England. These troops were to receive English pay.

It was further contemplated to land a British force of 10,000 men at St. Malo, who, in conjunction with the Royalists from Brittany, were to advance on Paris. This proposition was, however, never carried out, and the troops, although concentrated in the south of England with a view to embarkation, were never actually put on board.

Quiberon Bay was the point selected for disembarking the expedition as being a capacious and safe harbour, with a good beach for landing troops and stores, and also, as all the country in its immediate vicinity was affected to the Royalist cause. **1795. Expedition to Quiberon Bay**

The strength of the force which it was proposed to send was small; but it was hoped that the representations of the *emigrés* would prove true, and that the landing of such an expedition would cause a rising among the peasants, and stir up afresh the smouldering embers of resistance amongst the Chouans and Vendéans. With a view to this contemplated rising of the peasantry, the expedition was to carry with it sufficient arms, clothing, and stores to equip large bodies of volunteers.

About the middle of June, 1795, the expedition was embarked at Portsmouth on board fifty transports, under the convoy of Sir John Borlase Warren's squadron, consisting of three line-of-battle ships and six frigates.

The strength of the force embarked was about 2,500 men, of which the following are the approximate details:—

Four regiments of infantry, averaging each 300 men, viz., those of d'Hervilly, Dudresnay, d'Hector, and La Châtre; a regiment of artillery, under Rotalier, of 700 men; 60 field guns; 80 officers to command volunteers, surgeons, paymasters, intendants, &c. The Archbishop of Dôl and fifty priests also accompanied the expedition.

This force was termed the first division, a second division being in course of organisation in the island of Jersey, recruited during the previous year from Royalist *emigrés* in Germany.

The expedition was under a sort of dual leadership, the Comte de Puisaye being considered by one portion as its leader, whilst the Comte d'Hervilly, who held a commission from the King of England to command those troops on English pay, was held by another party to be the Commander-in-Chief. On the landing of the expedition this rivalry between de Puisaye and d'Hervilly broke out into an open quarrel, which an attempt was made to settle by referring the question back to England, but the answer to this inquiry did not arrive until after the collapse of the enterprise, when d'Hervilly was dead and de Puisaye was again on board ship.

Until the arrival of the answer from England, d'Hervilly occupied the position of chief; but this arrangement only partially answered, and the mutual jealousy displayed by the rival commanders was one of the principal causes of the unhappy failure of the expedition.

In addition to the troops carried, the ships brought with them 80,000 stand of arms, clothing for 60,000 men, saddlery, horses, supplies of food, wine, brandy, &c., an enormous quantity of shoes, several tons of gunpowder, and about 2,000,000*l.* sterling in specie.

On the 13th June, 1795, the expedition set sail, being accompanied until the 19th, when it had arrived off Belleisle, by the channel fleet, under command of Lord Bridport. On this day, the wind being fair for Quiberon Bay, the fleet stood out from the coast to keep a look out for the Republican fleet, which was known to have left Brest harbour, whilst the expedition continued on its

1795. Expedition to Quiberon Bay.

road to Quiberon. Not long after the departure of the Channel fleet the Republican ships came in sight of the expeditionary fleet, which immediately altered its course, steering in the direction taken by Lord Bridport, and sending a fast sailing frigate to acquaint him of the proximity of the Republican squadron. The latter must apparently have mistaken Sir John Warren and his convoy for the Channel fleet, as no attempt was made to capture any of the ships of the expedition. On the morning of the 20th June, Sir J. Warren came in sight of Lord Bridport, who altered his course so as to interpose between the ships of the expedition and the Republicans, and the two fleets proceeded in this manner until the morning of the 23rd, when they came into collision off the island of Groix, the result being that the Republicans took refuge, with the loss of three vessels, between that island and Port Lorient, thus opening the way for the expedition to proceed to Quiberon Bay. The defeat of the Republicans was not, however, an unmixed good for the Royalist cause, as the presence of the fleet at Lorient prevented the inhabitants of the town from siding as they wished, with the Royalist cause and joining in the general rising of the peasants. On the morning of the 25th the expedition entered Quiberon Bay, and the Comte de Puisaye, accompanied by several officers, landed and went into the interior to ascertain the state of affairs. At daybreak on the 27th, all the troops of the expedition disembarked at the village of Carnac. The landing was effected without any loss, the only opposition encountered being the harmless fire of some 250 Republican troops, a portion of the garrison of Auray, who were forced to retire with the loss of several killed, wounded, and some prisoners, before a large party of armed Chouans, who had come down to the coast at the sight of the ships of the expedition.

The troops on landing were placed in cantonments about the village of Carnac. More than a week elapsed before the Royalists attempted to effect anything beyond arming and organising the Chouans, who flocked to the raising of the Royal Standard to the number of some 14,000, of all ages. The Republicans at first retired at all points, leaving the country open from the coast to some thirty or forty miles inland; but the Royalists took no advantage of this and remained content with entrenching themselves at and about Carnac. In the interior of Brittany a large force of Chouans assembled under la Bourdonnaye, having been clothed and armed with stores landed from the ships; but they never effected a junction with the Royalists of the expedition, as, long before the latter had determined on their course of action, the Republicans, under General Hoche, having recovered from their first surprise, had placed their forces so as to prevent any such combination. On the 4th July, the Royalist forces about Quiberon Bay were disposed as follows:—the troops under d'Hervilly in and about the village of Carnac; the volunteers organised in three divisions: the first, under the Comte Dubois-Berthelot, about four miles from the coast of the Quiberon peninsula on the Auray road; the second, under General Tintigniac, about three miles north-west of the first, in front of the village of Landevant; and the third under the Comte de Vauban (who also commanded the whole of the volunteers), about the village of Mendon, slightly in rear of the other two, as a support to either in case of an attack from the direction of Vannes.

1795. Expedition to Quiberon Bay.

At this time d'Hervilly projected an attack against Fort Penthièvre, which guards the spit of land joining the peninsula of Quiberon to the mainland. To do this was a waste of time, and indeed was unnecessary, as the fort did not command the peninsula itself, and its garrison was only some 500 or 600 men; so that the magazines could have been established without first taking the fort, which, if the Royalists were successful in their advance into the interior, must have fallen into their hands.

On the morning of the 5th July some 3,000 Republicans, who had been concentrated at Vannes, advanced to attack the Royalist outposts at Pontsol, about three miles in advance of Auray, in the direction of Vannes. The Royalists received the attack with a well-sustained fire, but on the Republicans bringing up a field piece they retired behind the town of Auray, and carried on a desultory fight till evening, when the Republicans retired on Pontsol.

The same day an attack was made on Fort Penthièvre, which a few days before had been fired on by the ships of the squadron without effect. The attack, however, was a failure.

On the 6th, the Royalist position, being considered too much advanced, was withdrawn nearer the sea to the villages Kercado, Plouharnel, Carnac, and St. Barbe: and an attack was made on the Republican outpost in front of St. Barbe, but without effecting anything.

The same day about 2,000 Royalists were embarked at Carnac in the boats of the squadron, and landed together with 300 British marines, on the peninsula of Quiberon, to attack Fort Penthièvre from the south, whilst d'Hervilly at the head of some 3,000 Royalists, attacked it from the mainland. The garrison of the fort surrendered, a portion volunteering for the Royalist side, and the remainder being taken on board the ships of the squadron to escape the vengeance of the Chouans. After the capture of the fort the Royalists commenced making magazines for stores, food, &c., on the peninsula. The Republicans made an advance on the Royalist position, but were kept back by the fire of the fort and some armed boats of the squadron.

The Republicans again attacked the Royalist position, and after some fighting drove them in at all points. Some 18,000 men, women, and children, who had come inside the Royalist lines to avoid the Republican cruelties, were hurried in terrible confusion on to the peninsula along the narrow spit; hundreds of them, in their endeavours to escape the Republican fire, leaping over the palisades into the covered way of the fort.

At this period there were on the peninsula, some six miles long by three at its greatest breadth, 30,000 souls, who, from their numbers, could not all be properly fed, the fighting men getting half a ration, and the women and children about four ounces of rice.

Partly with a view to relieve the strain on the commissariat, and partly to effect a diversion further down the coast, General

1795. Expedition to Quiberon Bay.

Tintigniac, at the head of 8,000 Royalists, was embarked on board the boats of the squadron, and landed near Sarzeau. Advancing into the country he overthrew all opposition, burning and destroying villages; and penetrating as far as the environs of Vannes, he defeated a Republican battalion at the Château of Coetlogon, where, in the moment of victory, he was shot by a Republican grenadier. The Royalists, under Georges Cadoudal, who succeeded Tintigniac, did not succeed in advancing any further, but had to retire before superior forces.

During the next few days a great number of Chouans were taken on board the ships of the squadron, and disembarked on various points of the coast of Brittany, so that not more than some 5,000 souls were left on the peninsula.

About this period (10th July) the Republicans, who had been gradually concentrating under General Hoche, were disposed as follows:—An entrenched camp had been formed a little to the south of St. Barbe, which was strongly garrisoned and heavily armed. The rear of the Republican army was covered by a division under General Meunier, who was cantonned about Ploermel. Guarding the left was a brigade under General Laviolas, who occupied Kercado and the position near Carnac lately evacuated by the Royalists. The Republicans mustered about 10,000 men of all arms. On the 15th the second Royalist division, under the Comte de Sombreuil, arrived in Quiberon Bay. This force numbered about 1,000 men, and consisted of the infantry regiments of Béon, Damas, Salur, Rohan, and Périgord. The vessels which brought this reinforcement carried also with them numerous supplies of all kinds for the use of the Royalists. The troops did not disembark until the 17th, when they were cantonned about the village of St. Julien.

The Comte d'Hervilly had determined on making an attack in force on the Republican outpost of St. Barbe, on the 16th July, in order to break through their line. Admiral Warren and the Comte de Puisaye tried to persuade Comte d'Hervilly to defer his project until the disembarkation of Sombreuil's division, but without success.

In order to draw away some of the Republican troops from the defence of St. Barbe, General Vauban, with some 800 Royalists, was taken on board the boats of the squadron and landed, together with 200 British marines, near Carnac, so as to threaten the Republican left.

At daybreak on the 16th, d'Hervilly, with about 3,000 Royalist troops, 18 field guns, and 600 Chouan volunteers, advanced to the attack of St. Barbe, having defiled from the peninsula to within gunshot of the entrenched camp during the night. The advance was made by wings, the right wing composed of d'Hector's regiment on the extreme right, that of Dudresnay at an interval of about 80 yards on the left of the first, the Chouan volunteers being some distance in rear of the centre of the other two regiments. The left wing was formed of d'Hervilly's regiment alone. The regiments were in columns of half-companies to facilitate moving over the broken ground. Partly thrown out as skirmishers, and partly in column of companies, the regiment of la Châtre preceded

# EXPEDITION TO QUIBERON BAY 1795.

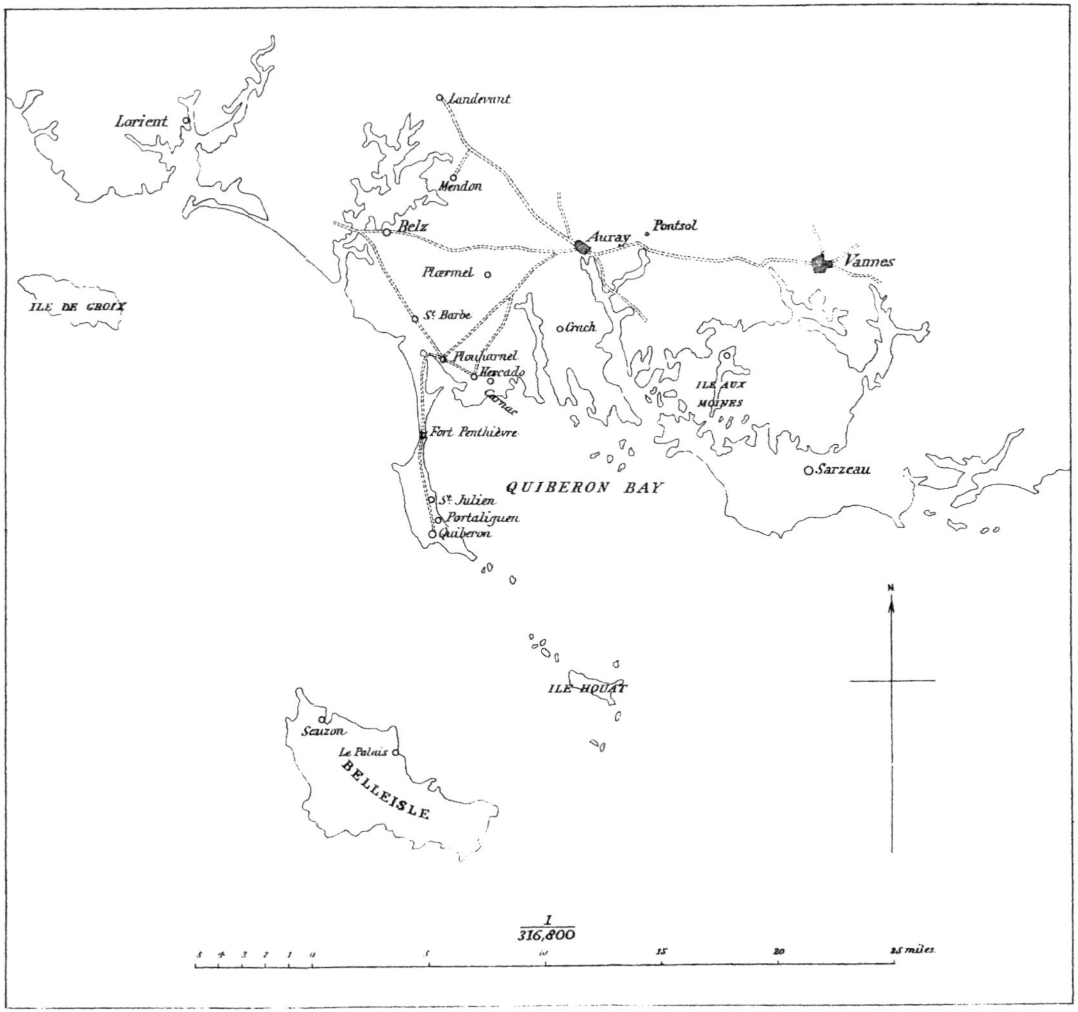

Lorient
ILE DE GROIX
Landevant
Mendon
Belz
Auray
Pontsol
Vannes
Plœrmel
St. Barbe
Crach
Plouharnel
Kercado
Carnac
ILE AUX MOINES
Fort Penthièvre
QUIBERON BAY
Sarzeau
St. Julien
Portaliguen
Quiberon
N
ILE HOUAT
Sauzon
Le Palais
BELLEISLE

1/316,800

5 4 3 2 1 0 5 10 15 20 25 miles.

Intelligence Branch N°304.

Litho.d at the Intelligence Branch War Office July 1883

1795. Expedition to Quiberon Bay.

the right wing. The artillery at first took up a position behind the advancing troops and dismounted some of the Republican guns by their fire; but being soon afterwards ordered to advance, got hopelessly fixed in the sandy ground in front of St. Barbe. On approaching the position of St. Barbe, d'Hervilly found that he was directing his attack on its strongest part; he therefore ordered the right wing to incline towards its left, thus exposing the regiments composing it to a front and enfilade fire from the Republican guns. The right wing was the first to come under fire, which was so heavy that d'Hervilly, before the left had become engaged, ordered it to retire to cover the retreat of the right wing, which appeared inevitable. After sustaining the Republican fire and making an attempt to charge, the right wing halted, turned, and retired in the greatest confusion, becoming in the rout mixed up with the left wing, and leaving five guns stuck in the sand, most of the teams having been shot down.

The force embarked in the boats of the squadron had failed in their attempt to land in front of Carnac, the beach being swept by some well-posted Republican guns. Failing in this attempt, they had turned towards the peninsula, and landed at the eastern side of the spit joining it to the mainland, just at the commencement of the rout of the other Royalists. By occupying the entrenchment at the head of the peninsula, General Vauban was enabled, with the aid of the fire from the armed boats of Sir John Warren's squadron, to prevent the Republican troops from entering the peninsula along with the flying Royalists. In this attack the Royalists lost heavily, especially in officers, among whom was the Comte d'Hervilly, who received a mortal wound, of which he died a few days later.

After this disaster there was a lull in the proceedings on both sides; but large numbers of the Royalists, chiefly French sailors who had been prisoners of war in England, deserted to the Republican side. Two of these deserters offered to guide two attacking columns along the strips of beach left at low tide on each side of the spit of land which united the peninsula with the mainland, and which were entirely concealed from the view of the fort on the spit.

Accordingly, the night of the 20th July having been selected for the enterprise, one column under General Humbert advanced along one side of the spit so as to get between the fort and the villages on the peninsula, where the main body of the Royalists were cantonned. Two other columns, one under Adjutant-General Ménage, the other under General Valletaux, were to attempt to capture the fort by escalade. A support under General Lemorne was left in front of the entrenched camp at St. Barbe.

The night was very dark and stormy, with heavy rain. The greater portion of the men who formed the garrison of the fort, belonging to d'Hervilly's regiment, were affected to the Republican side, and had furnished General Hoche with the countersign. On the advance of Ménage's column, some men of this regiment opened the gates of the fort and turned on their officers and that portion of the garrison which remained faithful to the Royalist cause.

1795. Expedition to Quiberon Bay.

To add to the general confusion most of the attacking column had been dressed in the red coats of the Royalists captured on the 16th July. By the time the two other columns had got on to the peninsula in rear of Fort Penthièvre, the Royalists had been alarmed and commenced a heavy fire on the approaching columns, who not being able to return the fire, owing to the priming of the muskets having become soaked with rain, began to waver and were about to turn, when, at the break of day, the Republican tricolour was seen waving over the fort. On seeing this the Royalists deserted their entrenchments and retired further on to the peninsula. The three Republican columns now united, advanced along the peninsula to attack the Royalists, who after a brief stand rushed down to the shore to await the boats of the squadron which were hastening to take them off. The regiment of Périgord, under the leadership of Sombreuil, fought with the most devoted valour to cover the embarkation of the Royalist fugitives, having established itself in the small redoubt of Portaliguen, which, however, was only strong on its sea face; but long before all the men, women, and children which composed the crowd of fugitives could be embarked, it had been forced to lay down its arms. An English corvette, the "Lark," succeeded in getting within gunshot of the land, and kept up a fire from her carronades on the Republican reinforcement hurrying to join in the slaughter which had commenced. Then ensued a scene the horror of which almost baffles description:—

"An immense crowd lined the shore, and raising their suppliant hands towards Heaven, implored from it that help which their fellow men refused them. Men, women, children, old men and soldiers, all heaped up in a struggling mass, waiting till the English boats should approach to save them from the Republican sword. But the tide was low and the shore bristling with reefs could not be approached. The Royalists rushed breast-high into the water to gain the rocks and reefs nearest the boats. Others more bold plunged into the sea to gain the boats swimming. The sailors were obliged to thrust back again those who clung to the boats with blows of oars and cutlass. The sea by this time having risen was covered with all sorts of wreckage, the remains of clothing and accoutrements, its foaming waves throwing up on the beach the bodies of those its waters had engulphed. The noise of the guns, the roar of the waves, joined to that of the peals of thunder, added new horror to the scene. The wretched Royalists, struggling for a foothold on the rocks, thrust one another back into the waves. Women advanced towards the boats, held aloft in their arms their babes, imploring in vain the pity of those surrounding them. This sentiment appeared to have become stifled in the hearts of all. . . ."

The gallant Sombreuil, with the remnants of the regiment of Périgord, trusting to the promise of quarter, laid down his arms: but the promise was treacherously broken. All Royalist prisoners were marched under escort to Auray and Vannes, where the Republican Commissions sat, and sentenced them, without exception, to be shot.

1795. Expedition to Quiberon Bay.

As a refinement of cruelty the procession of death was headed by a company of gravediggers, who proceeded to dig trenches for the bodies of the unfortunates who were kept watching the operations. All men over eighteen years of age, the gallant Sombreuil, the venerable Archbishop of Dôl, and fourteen priests were shot in a field near Vannes, now fitly known as the "Field of Martyrs."

On the morning of the 20th July there were about 4,000 souls on the peninsula; of these about 1,500 only were taken off by the ships of the squadron. Immense quantities of stores of every description fell into the hands of the Republicans; amongst other items, 10,000 stand of arms, 150,000 pairs of shoes, clothing for 40,000 men, in addition to which six transports, which had arrived on the evening of the 20th, laden with rum, brandy, and provisions, also became the booty of the Republicans.

The remnants of the Royalists were landed at the small island of Houat. Soon after landing some 700 were carried off by an epidemic fever.

The arms, &c., brought by the ships which had carried Sombreuil's division, had not been landed on the peninsula, and were a few days later landed at St. Jean de Mons, opposite the Ile d'Yeu, where they were received by a body of Chouans under Charette.

After a time the Royalists on the island of Houat gradually dispersed, and were landed, with the exception of some of the principal officers, on various points of the coast of Brittany.

In the beginning of October, 4,000 troops, of whom 2,000 were British under Major-General Doyle, arrived off Ile d'Yeu, the Comte d'Artois accompanying the expedition on board the "Jason." The troops were landed on Ile d'Yeu, together with large supplies of various stores. Here the force remained until the end of the year, when, having suffered considerably from sickness, and having lost all its horses, it was re-embarked at the end of December and returned to England.

## EXPEDITION TO OSTEND, 1798.

1798. Expedition to Ostend.

At the beginning of the year 1798, the greater part of the army destined for the invasion of England had been sent to Egypt under General Buonaparte, but a sufficiently large force still remained on the north coast of France to cause great anxiety in England. It was also ascertained that while numbers of light-draught boats were being built at Antwerp and Flushing, the canals leading to Ostend and Dunkirk were being enlarged to admit of their passage.

At Slykens, near Ostend, large sluice gates had been constructed, which were unprotected by any special work, and the English General commanding the southern district (Kent and Sussex), on the recommendation of Captain Popham, of the Royal Navy,

1798. Expedition to Ostend.

pointed out to the Government the facility with which these important gates might be destroyed.

The idea met with approval, and Sir Charles Grey was ordered to prepare an expeditionary force to be drawn from his command. Directions were also sent to the naval agent at Woolwich to provide the necessary stores; and boxes for petards were ordered to be made on board the ships of the navy in the Downs.

Sir Charles Grey selected the 11th Regiment and detachments from the 23rd and 49th Regiments and Royal Artillery, and placed them under the command of Major-General Eyre Coote. Portions only of the 23rd and 49th Regiments could be taken, owing to the number of Dutchmen in the ranks of the first, and of recruits in those of the second. The flank companies of the Guards, and a serjeant's party from the 17th Light Dragoons, were added to General Coote's command, which embarked and assembled off the North Foreland on the afternoon of the 15th May. The ships-of-war and transports were under the command af Captain Popham.

The expedition was directed against Ostend, with orders to destroy any shipping that might be at Blakenburg, and to make an attack on Flushing. It was thought, at the time, advisable to blockade the Texel, and to do so it was necessary to occupy the Isle of Ameland. The Dutch were to seize the island and hand it over to the English, who would hold it with the 11th Regiment, which was to be detached for that purpose from General Coote's force.

The fleet sailed at 4 A.M. on the 16th May, and anchored off Ostend at 1 A.M. on the 19th. At this time the weather was so threatening that the General and Naval Commander deliberated on the advisability of landing; meanwhile a pilot boat was brought alongside, and from the information given by the crew, General Coote learnt that Ostend was only held by a small force. This settled the question, and a landing was at all hazards immediately decided on. The disembarkation began about 3 A.M. on the sand-hills east of Ostend harbour, and many of the troops were ashore before they were discovered.

At 4.30 A.M., 19th May, the batteries and ships engaged, and in a short time Ostend was on fire. By 5 A.M. the troops, strength as below, were on shore:—

Serjeant's party, 17th Light Dragoons.
2 companies Light Infantry, Coldstream Guards.
2 „ „ „ 3rd Guards.
11th Regiment.
23rd „ flank companies only.
49th „ „ „
6 guns.

A party of seamen, under the direction of Captains Winthrop, Bradley, and McKellar, and Lieutenant Brady, of the Royal Navy, landed with tools, powder, &c.

The ships carrying that portion of the 23rd Regiment which had not disembarked were stationed to the west of Ostend to make a demonstration on that side, and, if opportunity offered, to land and spike the guns on the town works. These ships hove to at the

# EXPEDITION TO OSTEND 1798.

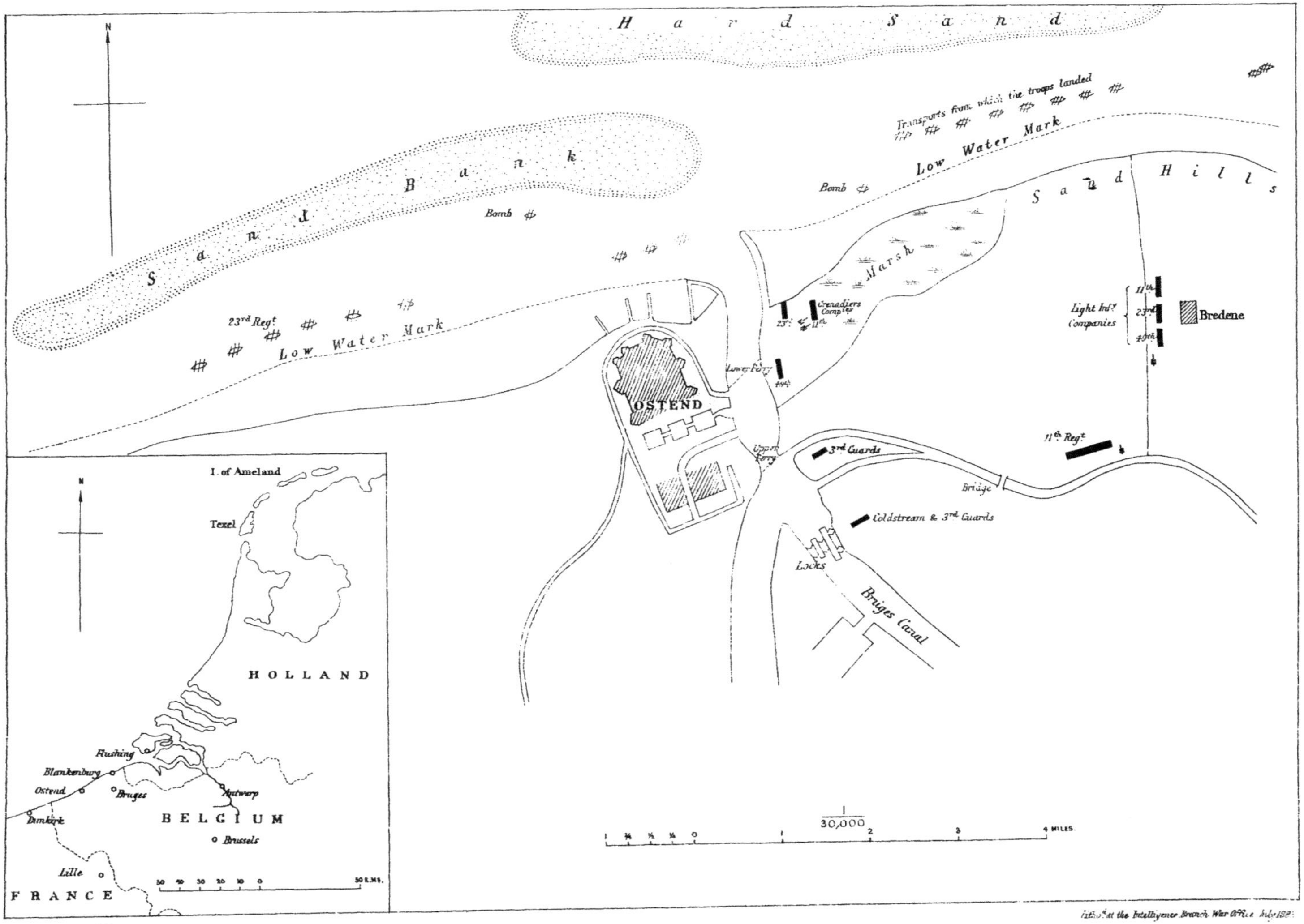

Intelligence Branch No. 307.

Litho.d at the Intelligence Branch War Office July 1883

**1798. Expedition to Ostend.**

mouth of the harbour, within 300 yards of the battery, and kept up such a heavy fire that the enemy's fifteen guns were silenced.

The Guards and Grenadier companies, under command of Major-General Burrard, with two 6-pounders, were directed to proceed to the harbour. In carrying out this order they met with some opposition from the enemy's sharpshooters, but eventually took up their positions. The Grenadiers of the 11th and 23rd Regiments with the guns, were posted at the Lower Ferry, to prevent the enemy crossing the harbour from Ostend. A detachment of Colonel Campbell's company of the Guards and the Grenadiers of the 49th Regiment, were posted at the Upper Ferry for a similar purpose.

The remainder of Colonel Campbell's company and the three other companies of the Guards carried out the operations at the head of the harbour, where the sailors, under the superintendence of Lieutenant Brownrigg, of the Royal Engineers, were engaged in destroying the sluice gates of the Bruges Canal.

The 11th Regiment was posted along a ditch to the south-east to cover the retreat.

The Light companies of the 11th, 23rd, and 49th Regiments, under Major Donkin, occupied the village of Bredene, and extended to the Blakenburg road near the sea.

At 9.30 A.M. the "Minerva" came up, but the sea was too rough to allow of the 1st Guards being landed, though Lieutenant-Colonel Warde put his men into boats intending to make the attempt, but was stopped by Captain Popham.

A summons to surrender having been sent into Ostend, the Commandant Muscar, although he had only 400 men, returned an indignant refusal.

By 10.30 A.M. the mines at the canal were completed; the explosion was successful in destroying the locks and sluices. Several vessels in the harbour were also destroyed.

Up to this time the English loss had only been five killed and wounded. The destruction of Bruges Canal gates rendered inland navigation from Flushing to Ostend impossible; thus the object of the expedition had been accomplished, and as the troops had landed with canteens only, and one day's cooked provisions in their haversacks, it was decided to re-embark. The force regained the beach by 11 A.M., and the seamen under Captain Winthrop succeeded in reaching their ships; but the increasing surf and wind soon prevented any further communication with the fleet.

General Coote, finding his retreat cut off, took up a position on the sandhills with his back to the sea, and had some breast-works constructed under the superintendence of Lieutenant Brownrigg, R.E.

The French on their side had not been idle: troops were hurriedly assembled from Bruges and the neighbouring villages. Kellar, the Commandant at Bruges, took command of the force thus collected, which, according to French accounts, never exceeded 500 men, viz.:—

1798. Expedition to Ostend.

250 men, 46th demi-brigade.
100 „ 94th „
150 „ of various corps.

But the garrison of Ostend is not included in the above, and it is more than probable that it lent a very considerable aid.

At 4 A.M., on 20th May, the English observed two columns advancing from the south, and soon afterwards other troops were seen on the flanks. The action was begun by the French Horse Artillery, whose fire was immediately replied to by the English guns. The contest lasted for two hours, until both flanks of the position were turned by the overwhelming numbers of the French. The 11th Regiment, which was posted on the left, having lost its Colonel (Haly), began to waver, and General Coote, in his endeavours to rally the men, was severely wounded. The command then devolved on General Burrard, who seeing the critical position of the force, ordered the cessation of fire and surrendered.

The artillery behaved splendidly, the officers working their guns to the last moment, and then, having spiked them, threw them over the bank into the ditch which had marked the front of the position.

Captain Popham had witnessed the action from his ship, but owing to the heavy sea which was running had been unable to afford any assistance.

About 11 o'clock A.M., seeing the English surrounded, he managed to get a flag of truce on shore, and the bearer returned immediately to inform him of the surrender.

According to returns drawn up by the Major of Brigade, Captain Thorley, the number of troops engaged was as follows:—*

2 Generals.
70 officers.
116 non-commissioned officers and drummers.
1,101 rank and file.
2 officers Royal Navy.
14 seamen.

Of these 1 General (Major-General Eyre Coote), 6 other officers, 2 officers Royal Navy, 6 non-commissioned officers and drummers, 135 rank and file, and 14 seamen were killed, wounded, or missing after the action.

The remainder, viz., 1 General (Major-General Burrard), 64 officers, 110 non-commissioned officers and drummers, and 966 rank and file surrendered unwounded.

The English wounded were taken into Ostend and attended by French surgeons. The remainder of the prisoners were removed to Lille.

As the object of the expedition had been attained the surrender did not cause the usual outcry against the officer commanding, and

* 17th Light Dragoons.
Brigade of Guards.
11th Regiment.
23rd Regiment.
49th Regiment.
Royal Artillery.
Royal Engineers.

in announcing the facts of the case to Parliament, the Government rendered every possible justice to the conduct of General Eyre Coote and the troops under his command.

1798. Expedition to Ostend.

---

## COMBINED BRITISH AND RUSSIAN EXPEDITION TO THE HELDER IN 1799.

1799. Expedition to the Helder.

THE arrogant determination of the French to force upon the other nations of Europe a Republican form of government had been so far successful that by the beginning of the year 1799, six minor Republics, viz., the Batavian, the Cisalpine, the Ligurian, the Helvetic, the Roman, and the Parthenopeian, were in subjection to France.

At that period the French forces, including auxiliaries, amounted to about 400,000 men, divided into armies of various strengths, carrying on operations in Holland, Germany, Italy, and Egypt, and closely observed by the troops of the Great Continental Powers, who only awaited an opportunity for again striking a blow at the common enemy.

For some time past Russia, Austria, and Prussia had been preparing for war; and the propitious moment seemed to have arrived when the Republican armies were so widely scattered as to be incapable of united action, and when England, under the direction of Mr. Pitt, was ready and eager to assist with men and money.

The second coalition against France was accordingly formed in January, 1799, and hostilities at once commenced. Considering the strength of the allied armies it was not surprising that at first the Republicans were unsuccessful.

In Germany, Jourdain and Soult were met by the Archduke Charles, who signally defeated them at Stockach, and compelled them to recross the Rhine. In Italy the veteran Russian General Suwarrow, victorious at La Trebia and Novi, had driven Macdonald and Joubert out of the fortresses of the quadrilateral, and from Milan and Turin. In Syria, Buonaparte had been foiled in his attempt to capture the port of St. Jean d'Acre by the Turks, assisted by the crews of two British ships-of-war under Sir Sydney Smith.

During the first months of the year England took no very active part in the struggle, but in June, with a view to further harass the enemy, it was decided to land an army of English and Russians in Holland. The Batavian Republic, formed of the states of Holland and Belgium, was well suited from its situation for the purposes of the expedition. It possessed an extensive seaboard, in which good natural harbours existed, and its eastern frontier was open to attack from the provinces of Hanover and Westphalia. The northern part (Holland), from the numerous canals intersecting it, was less suited to the movements of an army than the southern (Belgium), but had the advantage of being further removed from France. The chief fortresses, Amsterdam, Haarlem, Utrecht, Arnheim, Bois-le-duc, Bergen-op-Zoom, and Antwerp, were held by

1799. Expedition to Helder.

mixed garrisons of French, Dutch, and Belgian troops, and were mostly situated in Holland.

The objects of the expedition, besides the obvious one of forcing the French to concentrate for the defence of their hitherto unthreatened frontier, were the capture of the above-mentioned arsenals and the Dutch fleet, and the re-establishment of the power of the Stadtholder.

It was considered probable that the people of Holland, aware of the severe reverses experienced by the confederated Republics in other parts, and being directly threatened themselves, would either join the Allies or hold aloof.

By a treaty concluded with Russia on the 22nd of June, 1799, it was settled that that country should furnish 17,000 men to be paid by England. Two camps of 12,000 men each were formed at Southampton and in Kent, whence about 18,000 proceeded to the seat of war, making in all an army of about 35,000 men.

The English were commanded by His Royal Highness the Duke of York, Lieutenant-General Sir Ralph Abercromby acting as second in command, while the following General officers had Brigade commands, or were employed on the Staff :—

Lieutenant-General Dundas.
„ „ Sir J. Pulteney.
„ „ Hulse.
Major-General the Earl of Cavan.
„ „ the Earl of Chatham.
„ „ D'Oyley.
„ „ Manners.
„ „ Burrard.
„ „ Manners.
„ „ Sir J. Moore.
„ „ Hutchinson.
„ „ Don.
„ „ Coote.
„ „ H.R.H. Prince William of Gloucester.
„ „ Knox.

The Russian commanders were General Emme, Lieutenant-Generals Gerebzoff and Hermann, Major-Generals Essen, Sedmoratzky, Southoff, and Schutoff.

The numbers of the French and Dutch contingents have been variously estimated, but it may be taken that there were from 18,000 to 20,000 troops belonging to the Batavian Republic, and 8,000 French under Generals Brune, Daendels, Dumonceau, Vandamme, Boutet, and Van Guenicke.

By the beginning of August it had been decided to attempt a landing in North Holland near the Helder, which if successful would secure a safe harbour for the fleet, and place the army within striking distance of Amsterdam.

The first division of troops embarked on the 12th of August, and on the following day the convoy sailed for the low countries. Violent south-west gales delayed the transports for some time, but on the 20th the fleet stood in towards the Dutch coast, and every preparation was made to land on the 22nd. Unfortunately the weather again proved unfavourable, and it was not until the 26th August that the fleet came to anchor off the Helder, the troops disembarking at dawn on the 27th. 1799. Expedition to the Helder.

Had the English been able to land on the 22nd, there is every reason to suppose that they might have obtained possession of the peninsula of North Holland without striking a blow. At that time the French had their main force on the side of Zealand, with strong detachments in Groevingen and Friesland, so that they could only have brought up about 12,000 men in all, to oppose the landing. The appearance of the fleet off Texel, however, indicated the probable point of attack, and gave time to the enemy to concentrate for its defence.

The disembarkation was unopposed, but the advanced guard, under Sir J. Pulteney, had hardly begun to move forward when an action commenced which lasted until 3 P.M. General Daendel's division was posted near Callantsoog, a village on the coast some eight miles south of Helder, from which point he made repeated attacks on the right flank of the British, who, finding the country in their front occupied, had taken up a position in the sandhills, running north and south.

Although unable to extend more than one battalion for the protection of the threatened flank, the 23rd and 55th Regiments, under Colonel Macdonald, stubbornly held their ground until relieved by Major-General Coote's brigade. At last, about 3 P.M., finding that they could make no impression on the invaders, and having suffered severely from the fire from the fleet, the Dutch retired in the evening to Petten, a position situated about two leagues further south.

The disembarkation was rapidly carried on, and on the 28th August the whole of the troops were on shore. The immediate results of the action of the preceding day were: the possession of the Helder, from which the garrison had withdrawn during the night, and the capture of two line-of-battle ships, five frigates, and about thirteen Indiamen.

The entrance to the Zuyder Zee being now open to the British, the Dutch Admiral Story removed the fleet into the Vlieter, where it was followed by Admiral Mitchell on the 30th, and summoned to surrender. The crews of the Dutch ships having refused to fight, nothing remained for the Republican Admiral but to strike his flag. By this bloodless victory, six line-of-battle ships and an equal number of frigates fell into our hands.

From the 27th of August to the 1st of September the English remained in the immediate vicinity of the Helder, when the difficulty of supplying the troops, owing to the want of horses, compelled them to move forward to a position extending from Petten to Oude Sluis, with its front protected by the Zuype Canal. The French, to the number of 6,000, occupied Alkmaar, Bergen, and Egmond, with 10,000 Dutch troops on their right at Schermerhorn, and Avenhorn.

The country between the opposing forces was so much intersected with ditches and canals, except on the right opposite Petten, that Sir Ralph Abercromby would not risk an action until reinforced

1799. Expedition to the Helder.

either from England or the Baltic, and employed the weary time of waiting in strengthening his line.

General Brune, aware of the reasons for this unusual delay on the part of his adversary, determined to take advantage of them, and on the 10th September, at daybreak, attacked with his whole force. The right column, under General Daendels, moved against the village St. Maarten; the centre, under General Dumonceau, advanced on Krabbendam; while the left, under General Vandamme, made an attack on Petten, which was defended by the two brigades of Guards under Major-General Burrard. The attempts made to force these points were gallant but unavailing, owing to the strength of the position and the determination of the defenders, who suffered comparatively little, whereas the Republican losses amounted to from 800 to 1,000, killed and wounded. About 10 A.M. the enemy fell back on Alkmaar, pursued for some distance by the reserve under Colonel Macdonald.

The reasons which had determined Sir Ralph Abercromby to remain on the defensive still existed, and the recent losses sustained by the enemy were not sufficient to warrant a departure from the plan adopted. It was known that the Duke of York, with a large body of troops, had left England, and the Russian contingent was hourly expected; it was therefore considered advisable to undertake no offensive operations until after the concentration of the army.

Between the 12th and 15th September three Brigades of British troops, and two of the Russian Divisions, under Lieutenant-General Hermann and Major-General Essen, disembarked at the Helder, and proceeded to the advanced position on the Zuype Canal. The Duke of York landed on the 15th, and assumed the command of the army, which now amounted to about 30,000 men, with 1,200 Light Cavalry.

At this period the Allies possessed a superiority of force with which it was material, as early as possible, to strike a decisive blow. The Dutch, to the number of 12,000, were in a strong position about Langedijk, somewhat in advance of the French, who, by drawing in all detachments, had raised their field strength to 10,000 men, who were posted in Alkmaar, Bergen, Schoorl, and Egmond-aan-Zee.

On the 19th of September the forces, under the Duke of York, formed in four columns, moved forward from Schagenbrug. The left column, commanded by Sir Ralph Abercromby, consisted of—

Two squadrons 18th Light Dragoons.
Major-General the Earl of Chatham's Brigade.
Major-General Sir J. Moore's Brigade.
Major-General the Earl of Cavan's Brigade.
First battalion of British Grenadiers of the line.
First battalion of Light Infantry of the line.
The 53rd and 55th Regiments, under Colonel Macdonald.

The right column, commanded by Lieutenant-General Hermann, consisted of—

The 7th Light Dragoons.

Twelve battalions of Russians. 1799.
Major-General Manners's Brigade. Expedition to the Helder.

The left-centre column, commanded by Lieutenant-General Sir James Pulteney, was composed of—

Two squadrons 11th Light Dragoons.
Major-General Don's Brigade.
Major-General Coote's Brigade.

The right-centre column, under Lieutenant-General Dundas, consisted of—

Two squadrons 11th Light Dragoons.
Two brigades of Foot Guards.
Major-General H.R.H. Prince William's Brigade.

The plan of operations was as follows :—The left column was to turn the enemy's right, on the Zuyder Zee; the right was to drive the enemy from the heights of Camperdown, and to seize Bergen; the right-centre had to force the position at Warmenhuizen and Schoorldam, and to co-operate with the right column; while the left-centre had to obtain possession of Oudkarspel, on the main road leading to Alkmaar.

The enemy's left was advantageously posted on the high sandhills which extend from the sea, in front of Petten, to the town of Bergen. The ground over which the centre columns had to move was intersected every three or four hundred yards by broad, deep, wet ditches and canals. The bridges across the few roads leading to the points of attack were destroyed, and obstacles had been carefully arranged.

The action was commenced by Lieutenant-General Hermann's column, which had by 8 A.M., 19th September, obtained possession of Bergen; but in moving against the main body of the enemy, posted in the woods surrounding the village, the Russian troops lost their order, and were driven back. Lieutenant-Generals Hermann and Tchertchekoff were made prisoners, and their troops were forced back through Bergen to Schoorl, which they also had to abandon. This village was immediately retaken by Major-General Manners's Brigade, which was then reinforced by two battalions of Russians, by Major-General D'Oyley's Brigade of Guards, and by the 35th Regiment, under Prince William. The action was renewed by these troops, who in their turn repulsed the enemy; but want of ammunition and the exhausted state of the corps engaged in that part of the field obliged them to retire on Petten and the Zuype Canal.

The column under Lieutenant-General Dundas, at dawn, attacked the village of Warmenhuizen, where the enemy, with a large force of artillery, was strongly posted. Three battalions of Russians, under Major-General Sidmoratzky, moving from Krabbendam, gallantly stormed the left of the village, the 1st Regiment of Guards entering it on the right at the same time. The Grenadier battalion of Guards, the 3rd Regiment of Guards, and the 2nd battalion 5th Regiment, which had been previously detached to march upon Schoorldam to keep up the communication with Sir James Pulteney, were joined by

1799. Expedition to the Helder.

the remainder of the column, which, after taking Warmenhuizen, had been reinforced by the 1st battalion 5th Regiment, and the whole moved forward and seized the village, which they held under a galling fire of artillery until the conclusion of the action.

The left-centre column, though opposed by the bulk of the Batavian army, under General Daendels, had overcome all opposition and taken possession of Oudkarspel, thus securing the direct line of advance on Alkmaar. Sir Ralph Abercromby had equally well accomplished his task by capturing the town of Hoorn, on the coast of the Zuyder Zee, and placing himself in a favourable position for completing the turning movement. However, in consequence of the partial failure on the right, it was considered necessary to recall all the troops and re-occupy the former position.

In this action the force under Sir Ralph Abercromby took no direct part; consequently the allied troops engaged amounted to no more than from 15,000 to 18,000 men.

The losses on both sides were considerable :—

English: 6 officers, 2 serjeants, 109 rank and file, killed; 43 officers, 20 serjeants, 2 drummers, 345 rank and file, wounded; 22 serjeants, 5 drummers, 463 rank and file, missing.

Russians: 1,741 non-commissioned officers, rank and file, and 44 officers, killed or captured. 1,225, including 49 officers, wounded.

Republicans: 3,000 prisoners, including 60 officers. 16 guns taken.

The failure of this attack by the Allies was due to the careless action of the Russian troops, and to the loss of their two principal officers at the most critical period of the struggle. The corps under Sir Ralph Abercromby began their march on the evening of the 18th September, but his advance was delayed by the bad state of the roads, and he arrived at Hoorn many hours later than was expected. The objects to be gained by the movements of this column would, if attained, have had a material effect on the result of the whole expedition, and could only be attempted while the Duke of York possessed a superiority of force. The enemy had left their right uncovered, and a very strong country unoccupied, from which it was evident that it would have been very difficult for them to correct their error if the attack on that flank had been vigorously carried out, and they had also left Amsterdam undefended on the only side by which it was accessible. The strength of the column which attacked Bergen would have been more than sufficient if it had been employed with common prudence. This column was at all times very superior in numbers to the enemy who opposed it, but it moved in mass in an intersected country, never covered its flanks, and its operations having, contrary to order, been commenced long before daylight, its fire was probably more destructive to itself than to the enemy.

That the other columns were not too weak is sufficiently shown by their having taken and held, until recalled, the points against which they were directed.

The Republicans re-occupied all the positions from which they had been driven, and their general line of defence was now covered on the right by inundations, the only roads across which were covered by field works. The space between Alkmaar and the

Zuyder Zee was thus rendered defensible by small numbers, and Amsterdam was secured on the land side. The remainder of the army, which had been reinforced, was concentrated between the Langedijk and the sea, and the post of Oudkarspel was strengthened by additional works, and by inundations. Schoorldam and the Koedijk were also fortified.

1799. Expedition to the Helder.

The Duke of York was anxious to renew the attack before the enemy should be further reinforced, and the arrival of the third Division of Russians, under General Emmé, on the 26th September, and of some detachments from England, had more than replaced the numbers lost in the last battle. The state of the weather and the roads, however, compelled him to defer operations until the 2nd of October.

It has already been mentioned that the right of the enemy was no longer assailable, and that the further precautions taken by them had rendered an attack in front unadvisable. His Royal Highness therefore determined to operate with his main force against their left, with the view of forcing them to evacuate North Holland.

The combined attack was made in four columns. The first on the right, under General Sir Ralph Abercromby, consisted of—

Major-General D'Oyley's Brigade.
Major-General Sir J. Moore's Brigade.
Major-General the Earl of Cavan's Brigade.
Colonel Macdonald's Reserve.
Nine squadrons of Light Dragoons, under Colonel Lord Paget.
One troop of Horse Artillery.

This column moved along the beach against Egmond-aan-Zee, with a view to turn the enemy's left flank.

The second column, commanded by Major-General Essen, consisting of Russian troops, moved by the Slaper Dyke, which skirts the sand hills of Camperdown, through the villages of Groet and Schoorl, upon Bergen, with its left flank covered by a detachment under Major-General Sedmoratzky, which was to support the Brigade under Major-General Burrard in the attack of Schoorldam.

The third column, under the command of Lieutenant-General Dundas, consisted of:—

Major-General the Earl of Chatham's Brigade.
Major-General Coote's Brigade.
Major-General Burrard's Brigade.
One squadron 11th Dragoons.

Major-General Coote's Brigade was to follow the advanced guard of the first column from Petten: to turn to the left at the village of Campe, and clear the road to Groet and the heights above that town, for that part of the Russian column which marched by the Slaper Dyke, whose right Major-General Coote was to cover during its advance on Bergen.

Major-General Lord Chatham's Brigade was to support the main portion of the Russian column. Major-General Burrard's Brigade was to move along the left of the Alkmaar Canal, and to combine with the corps under Major-General Sedmoratzky in its

1799. Expedition to the Helder.

attack on Schoorldam, communicating at the same time with the fourth column, which was advancing on his left, under Lieutenant-General Sir James Pulteney; this corps was composed of—

Major-General H.R.H. Prince William's Brigade.
Major-General Manners's Brigade.
Two squadrons 18th Light Dragoons.
Two battalions of Russians.

It covered the left of the Allies as far as the Zuyder Zee, and was to threaten the enemy's right, and take advantage of any favourable circumstance that might offer.

The enemy was supposed to have 25,000 men in the field, the greater number of whom were French.

The right column moved from Petten at 6.30 A.M., 2nd October. Its advanced guard, under Colonel Macdonald, drove the enemy from Campe and from the sandhills above that village, and moved forward, followed by the remainder of the division in the order above described. The enemy, who had gradually retired from Schoorl, were now formed in considerable force from Schoorl to Schoorldam, from which position they were driven by the Russians about 11 A.M. They retired on Bergen, and the ground between it and the Koedijk, which they held for the remainder of the day. Meanwhile the sandhills, on the left of the enemy's line, which were held by a large force, were attacked and cleared by the Brigades under Major-General Coote and the Earl of Chatham.

The Division under Sir Ralph Abercromby, meeting with but little resistance at first, had arrived within a mile of Egmond, where a large body of French infantry was found strongly posted on the sandhills in front of the town, with their left covered by cavalry and artillery on the beach. At this point the enemy was decidedly the stronger, and attacked with great determination; but every attempt to drive back the English was unsuccessful, and by nightfall they had pushed their advance to within a short distance of Egmond.

Lieutenant-General Sir James Pulteney had assembled the greater part of his column in front of Dirkshorn, whence he threatened Oudkarspel, in and near which was placed the principal force of the enemy's right, and effectually prevented him from sending any detachments to his left.

The result of this day's action was that early on the 3rd October the Republicans evacuated their strongly fortified posts at Oudkarspel and Langedijk, retiring upon St. Pancras and Alkmaar. They still held the town of Bergen, and appeared in some force on that side near Koedijk. The main body, however, had been withdrawn during the night, and before noon the village was occupied by the 35th Regiment. About 10 o'clock A.M., Sir Ralph Abercromby entered Egmond-aan-Zee, and in the evening the Russians advanced to Egmond-op-Hoef. At the same time Major-General Burrard, who when the enemy evacuated Bergen had moved up to Koedijk, sent forward a detachment to take possession of the town of Alkmaar, which had been abandoned. The result of the action of the 2nd October was that the enemy had been obliged to retire from his strong advanced position to Beverwijk,

# EXPEDITION TO THE HELDER 1799.

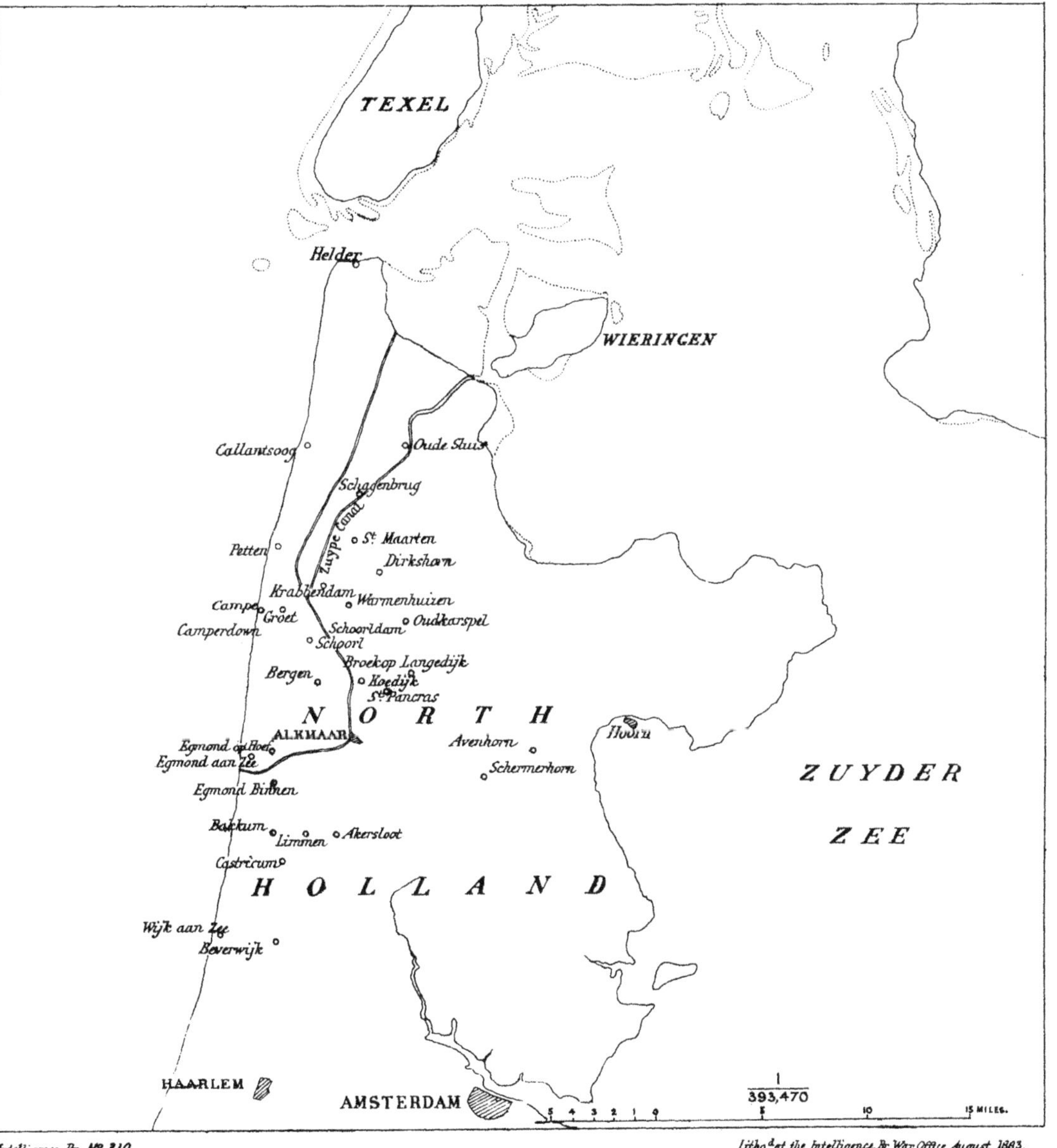




Litho.d at the Intelligence Br. War Office August 1883.

and Wijk-aan-Zee, which, owing to the exhausted state of the victors and the nature of the country, he was allowed to do in good order.

1799. Expedition to the Helder.

The fighting had been very severe, and the losses on both sides were heavy. The enemy was supposed to have lost 4,000 men, and seven guns had been captured.

The English lost—1 major, 5 captains, 5 subalterns, 11 serjeants, 215 rank and file, killed; 1 major-general, 2 colonels, 3 lieutenant-colonels, 4 majors, 23 captains, 40 subalterns, 47 serjeants, 7 drummers, 980 rank and file, wounded; 1 captain, 4 subalterns, 3 drummers, 178 rank and file, missing.

The Russian losses were—1 field officer, 1 captain, 3 subalterns, 9 non-commissioned officers, 157 privates, killed or taken prisoners; 1 General, 1 field officer, 1 captain, 18 subalterns, 9 non-commissioned officers, 365 privates, wounded.

On the 4th, the right of the army, under Sir Ralph Abercromby, pushed forward its outposts to beyond Egmond-aan-Zee, Egmond-op-Hoef, and Egmond Binnen; the centre occupied Alkmaar, and the villages in its front towards Limmen; the left was placed behind the canal of Alkmaar, between that town and Schermerhorn. The town of Hoorn, on the Zuyder Zee, was re-occupied by a detachment from the left. The roads were in a dreadful state, and the conveyance of supplies was very difficult; this circumstance, and the necessity of giving some repose to the troops, prevented the Duke of York from immediately following up the successes obtained on the 2nd and 3rd October. On the 6th, however, he determined to push forward the outposts of the centre and right, driving the enemy from the villages of Akersloot, Limmen, and Bakkum, and from the ground between that village and the sea.

It was intended that this advanced line should be held preparatory to the general attack in contemplation, but Colonel Macdonald having followed the enemy too eagerly on the right, and the Russians having advanced to Castricum, which it was not intended they should attack, brought on a contest between these corps and the reinforcements sent by the enemy, which, although it had no object, soon extended along the whole front. By the evening the Republicans had been driven back, and the Allies retained possession of Bakkum and Castricum.

The losses sustained in this useless action rendered its effects equivalent to a defeat, and compelled His Royal Highness to suspend the meditated attack at a moment when the Republicans were likely to acquire a numerical superiority. These circumstances, the approach of winter, and the unfavourable nature of the advanced line as a defensive position, added to the fact that the Dutch people had not risen in favour of the invaders, rendered it evident that no movement in advance could be attempted with any chance of success, nor could the army remain where it was.

Acting, therefore, on the advice of Sir R. Abercromby and the other Generals, the Duke of York decided to retired to the Zuype position, where he would be nearer to his base. The troops began to retire on the evening of the 7th October, and the centre and right

1799.
Expedition to the Helder.

reached the Zuype Canal on the following day. The left wing came into line on the 12th October, having been pressed during its march by the Republicans, who, however, did not move forward in any strength until the Allies were in position.

Henceforth the resumption of offensive operations was out of the question, and His Royal Highness had to decide between remaining on the defensive or of endeavouring to evacuate North Holland. The position which he occupied was good, and he might have resisted for a long time all attempts to force it; but at that season of the year the arrival of supplies and reinforcements could not be depended on, and sickness was on the increase. The enemy, on the contrary, were daily acquiring strength, and could attack at pleasure without fearing any deficiency of supply, and if repulsed their retreat was open. Whereas, in the event of the allied position being forced, the embarkation would have to be effected under the attack of a superior enemy, allowing even that the weather admitted of a sufficient number of transports being collected at the required moment.

These considerations, and the paramount duty of preserving his troops when no object could be obtained by incurring further loss, induced the Duke of York to enter into negotiations for the evacuation of the country. The agreement was concluded on the 18th of October, and from that day hostilities ceased. The Allies were to evacuate the territory and seas of Holland before the 30th of November, retaining, however, possession of the captured fleet.

The greater part of the troops were on board before the end of October. The Duke of York embarked on the 1st of November, leaving to Sir James Pulteney the final execution of the agreement; the latter left the Texel on the 19th November with the remaining troops, and the French re-occupied the Helder on the same day.

---

Appendix I.

## APPENDIX I.

Articles of Evacuation.

Articles agreed upon between Major-General Knox, duly authorised by His Royal Highness the Duke of York, Commander-in-Chief of the combined English and Russian Army, and Citizen Rostollan, General of Brigade and Adjutant-General, duly authorised by Citizen Brune, General and Commander-in-Chief of the French and Batavian Army.

ARTICLE I. From the date of this convention all hostilities shall cease between the two armies.

ARTICLE II. The line of demarcation between the said armies shall be the line of their respective outposts as they now exist.

ARTICLE III. The continuation of all works, offensive and defensive, shall be suspended on both sides, and no new ones shall be undertaken.

ARTICLE IV. The batteries taken possession of at the Helder, or at other positions within the line, now occupied by the combined English and Russian army, shall be restored in the state in which they were taken, or (in case of improvement) in their present state, and all the Dutch artillery therein shall be preserved.

ARTICLE V. The combined English and Russian army shall embark as soon as possible, and shall evacuate the territory, coasts, islands, and inland waters of the Dutch Republic by the 30th of November, 1799, without committing any injury by inundations, cutting the dykes, or otherwise interfering with the means of navigation.

1799. Expedition to the Helder. Articles of Evacuation.

ARTICLE VI. Any ships-of-war, or other vessels, which may arrive with reinforcements for the combined British and Russian army, shall not land the same, and shall be sent away as soon as possible.

ARTICLE VII. General Brune shall be at liberty to send an officer within the lines of the Zuype, and to the Helder, to report to him the state of the batteries and the progress of the embarkation. His Royal Highness the Duke of York shall be equally at liberty to send an officer within the French and Batavian lines, to satisfy himself that no new works are carried on on their side. An officer of rank and distinction shall be sent from each army respectively to guarantee the execution of this convention.

ARTICLE VIII. Eight thousand prisoners of war, French and Batavians, taken before the present campaign, and now detained in England, shall be restored without conditions to their respective countries. The proportion and the choice of such prisoners for each to be determined between the two Republics. Major-General Knox shall remain with the French army to guarantee the execution of this Article.

ARTICLE IX. The cartel agreed upon between the two armies for the exchange of the prisoners taken during the present campaign, shall continue in full force till it shall be carried into complete execution; and it is further agreed that the Dutch Admiral de Winter shall be considered as exchanged.

Concluded at Alkmaar, the 18th of October, 1799, by the undersigned General officers, furnished with full powers to this effect.

(Signed) J. KNOX, *Major-General.*
(Signed) ROSTOLLAN.

---

## APPENDIX II.

*Composition of the Franco-Dutch Army.*

Appendix II. Composition of Franco-Dutch Army.

### FRENCH ARMY.

French Army.

#### GENERAL STAFF.

General Commanding-in-Chief—Brune.
Chief of the Staff—General of Brigade Rostollan.
Commanding the Artillery—General of Brigade Saint-Martin.
Commanding the Engineers—General of Brigade Saint-Julien.
General of Division Vandamme.
" " Boudet.
" " Gouvion.
" " Barbou.
General of Brigade David.
" " Simon.
" " Fuzier.
" " Clément.
" " Pachtod.
" " Dardenne.
" " Aubrée.

1799.
Expedition to the Helder.
App. II.
(Composition of French Army.)

General of Brigade Durutte.
" " Dazémar.
" " Malher.
" " Paradis.
Adjutant-General Maison.
" " Delecourt.
" " Massabeau.
" " Aniel.

TROOPS.

*Infantry.*

| Demi-Brigades. | | | | Battalions. |
|---|---|---|---|---|
| 22nd | .. | .. | .. | 3 |
| 42nd | .. | .. | .. | 3 |
| 48th | .. | .. | .. | 1 |
| 49th | .. | .. | .. | 3 |
| 51st | .. | .. | .. | 3 |
| 54th | .. | .. | .. | 3 |
| 60th | .. | .. | .. | 1 |
| 72nd | .. | .. | .. | 2 |
| 90th | .. | .. | .. | 3 |
| 98th | .. | .. | .. | 2 |

*Cavalry.*

| Regiments. | | | Squadrons. |
|---|---|---|---|
| 10th Dragoons | .. | .. | 4 |
| 4th Chasseurs | .. | .. | 4 |
| 5th " | .. | .. | 1 |
| 16th " | .. | .. | 4 |

*Artillery.*

4th company of the 4th regiment Horse Artillery.
1st " " 8th " " "
2 companies of the 6th and 7th regiments Foot Artillery.

---

Composition of Dutch Army.

## DUTCH ARMY.

Lieutenant-General Daendels.
" " Dumonceau.
Major-General Van Guérik.
" Van-Zuylen-van-Nywelt.
" Bonhomme.
" Van-Boccop.
" Rietwelt.
" Bruce.
Adjutant-General Solivier.
" " Vichery.
" " Van Uslar.
" " Raaff.
Colonel Martus Chewitz, commanding Artillery.
Lieutenant-Colonel Krayenhoff, commanding Engineers.

1799. Expedition to the Helder. App. II. (Composition of Dutch Army.)

TROOPS.

*Infantry.*

| Demi-Brigades. | Battalions. |
|---|---|
| 1st .. .. .. .. .. | 3 |
| 2nd .. .. .. .. .. | 3 |
| 3rd .. .. .. .. .. | 1 |
| 4th .. .. .. .. .. | 2 |
| 5th .. .. .. .. .. | 3 |
| 6th .. .. .. .. .. | 3 |
| 7th .. .. .. .. .. | 3 |
| 1st battalion of Foot Chasseurs .. | 1 |
| 2nd " " " .. .. | 1 |
| 3rd " " " .. .. | 1 |
| 4th " " " .. .. | |

*Cavalry.*

| Regiments. | Squadrons. | |
|---|---|---|
| 1st .. .. .. | 4 | |
| 2nd .. .. .. | 2 | |
| Dragoons .. .. | — | |
| Hussars .. .. .. | 4 | |
| Battalions { French .. .. | 24 | 46 |
| Battalions { Batavian .. | 22 | |
| Squadrons { French .. .. | 15 | 29 |
| Squadrons { Batavian .. | 14 | |

---

*Composition of the Anglo-Russian Army.*

Composition of Anglo-Russian Army. English Army.

ENGLISH ARMY.

GENERAL STAFF.

Commander-in-Chief—H.R.H. the Duke of York.
Major-General Farringdon, commanding Artillery.
Colonel Hay, commanding Engineers.
Colonel Anstruther, Q.M.G.
" John Hope, A.A.G.
" Alexander Hope, A.A.G.
Lieutenant-General Ralph Abercromby.
" " Sir James Pulteney.
" " Dundas.
" " Hulse.
Major-General Doyle.
" Burrard.
" Coote.
" Moore.
" Don.
" Earl of Cavan.
" Earl Chatham.
" H.R.H. Prince William of Gloucester.
" Manners.
" Hutchinson.
" Knox.
Lieutenant-Colonel Macdonald, commanding Reserve.

**1799. Expedition to the Helder. App. II., Composition of English Army.**

TROOPS.

| Divisions. | Brigades. | Regiments. | Battalions. |
|---|---|---|---|
| First, Lt.-General Sir R. Abercromby. | 1st Guards, Doyle. | Grenadier Guards | 1 |
| | | 1st Guards .. | 1 |
| | 2nd Guards, Burrard. | 2nd " .. | 1 |
| | | 3rd " .. | 1 |
| | 1st Infantry Brigade, Coote. | 2nd Foot .. .. | 1 |
| | | 27th " .. .. | 1 |
| | | 29th " .. .. | 1 |
| | | 85th " .. .. | 1 |
| | 2nd Infantry Brigade, Moore. | 14th " .. .. | 1 |
| | | 21st " .. .. | 1 |
| | | 49th " .. .. | 1 |
| | | 79th " .. .. | 1 |
| | | 92nd " .. .. | 1 |
| Second, Lt.-General Sir J. Pulteney. | 3rd Brigade, Don. | 17th " .. .. | 2 |
| | | 40th " .. .. | 2 |
| | 4th Brigade, Lord Cavan. | 20th " .. .. | 2 |
| | | 63rd " .. .. | 1 |
| | 5th Brigade, Colonel Macdonald. | 23rd " .. .. | 1 |
| | | 55th " .. .. | 1 |
| Third, Lt.-General Dundas. | 6th Brigade, Lord Chatham. | 4th " .. .. | 3 |
| | | 31st " .. .. | 1 |
| | 7th Brigade, Prince William. | 5th " .. .. | 2 |
| | | 35th " .. .. | 2 |
| | 8th Brigade, Manners. | 9th " .. .. | 2 |
| | | 56th " .. .. | 1 |
| Advance Guard, Major-Genl. Knox. | Grenadiers of Line .. | .. .. .. | 1 |
| | Light Infantry of Line | .. .. .. | 1 |
| Artillery | 3rd Regiment. | | |
| | 4th " | | |
| | Horse Artillery, 1 troop. | | |

*Cavalry.*

| | | Squadrons. |
|---|---|---|
| 7th Light Dragoons | .. .. | 4 |
| 11th " " | .. .. | 4 |
| 15th " " | .. .. | 4 |
| 18th " " | .. .. | 2 |

---

**Russian Army.**

## RUSSIAN ARMY.

Lieutenant-General Hermann.
" " Jerepsoff.
Major-General Soudhoff.
" Capzewitz.
" Essen.
" Darbinioff.
" Sedmoratzky.
" Emmé.

TROOPS.

| Division. | Brigades. | Regiments. | Battalions. |
|---|---|---|---|
| First, Jerepsoff. | Colonel Count de Fersen | Regiment of Jerepsoff | .. 2 |
| | | Regiment of Fersen | .. 2 |
| | Colonel Doubiansky | Grenadier Regiment | .. 3 |

1799. Expedition to the Helder. App. II. Composition of Russian Army.

| Divisions. | Brigades. | Regiments. | Battalions. |
|---|---|---|---|
| Second, Essen. | Sedmoratzky .. | Regiment of Sedmoratzky | 2 |
| | | Grenadier Regiment .. | 2 |
| | Darbinioff .. .. | Regiment of Darbinioff.. | 2 |
| | | Grenadier Regiment .. | 1 |
| Third, Emmé. | Capzewitz .. .. | Regiment of Emmé .. | 2 |
| | | Grenadier regiment .. | 2 |
| Advance Guard | Sondhoff .. .. | Regiment of Chasseurs .. | 2 |
| | | Grenadiers of Pétersbourg | 1 |

| | | |
|---|---|---|
| Artillery .. | Regiment of Capzewitz .. | 1 battalion. |
| Cavalry .. | Regiment of Hussars .. | 4 squads. |
| | Cossacks, 7th Division .. | 6 „ |
| Battalions | English, 37; Russian, 22 .. .. | 59 battalions. |
| Squadrons | English, 14; Russian, 10 .. .. | 24 squadrons. |

## EXPEDITION TO COPENHAGEN, 1807.

1807. Expedition to Copenhagen.

WHEN the Battle of Trafalgar put an end to Napoleon's hopes for the invasion of England with the flotilla collected at Boulogne, he formed another project for carrying out his designs against this country. The first part of his plan was to combine all the continental states into one great alliance against England, and compel them to exclude the British flag and British merchandise from their harbours: this was his famous "Continental System." The second part of the plan was to obtain possession of all the fleets of Europe, and by their means effect a descent on the shores of Great Britain. He thus hoped to weaken the resources of England and to spread distress amongst its inhabitants, and then to make a successful attack upon her. In this view, it had been decided in the secret article of the Treaty of Tilsit that the navies of Denmark and Portugal were to be demanded from their respective sovereigns, and seized by force if not voluntarily surrendered. When the Treaty of Tilsit was made Napoleon was at the height of his power, the strength of Prussia was completely broken, Austria was overawed, and Russia had at last been defeated. No force on the Continent seemed capable of withstanding the French arms; and Sweden, which still opposed him was far removed from the scene of European strife, and would probably soon be subdued by the forces of the French or Russians. The king of that country, however, did not despair, with the assistance of England, of making head against his enemies. The advanced guard of an English force, which had been destined for the support of Russia and Prussia, appeared in the Baltic in July, 1807, and the King of Sweden then denounced the armistice which had been concluded, just nineteen days after the Battle of Friedland.

Napoleon speedily assembled 30,000 men under Marshal Brune, who on the recommencement of hostilities began on all sides to press the 15,000 Swedes in Pomerania.

The Swedes retired to Stralsund, which was completely invested

1807. Expedition to Copenhagen.

by Marshal Brune in the middle of July. Lord Cathcart, with 10,000 British troops, formed part of the garrison of this place until, at the end of the month, they were withdrawn to co-operate in the expedition against Copenhagen. Stralsund held out until the 20th August, when the Swedes withdrew to the island of Rugen, which latter they evacuated and gave over to the French troops on the 7th September.

The English Government obtained information of the secret article of the Treaty of Tilsit almost as soon as it had been framed. The French Emperor lost no time in acting upon it; Denmark was summoned to join him in an active alliance against England, and troops were marched into Holstein in such numbers that it was evident Denmark would lose all her continental possessions if she resisted the demands of the allied Emperors.

It was therefore absolutely necessary that the Cabinet of Great Britain should act with vigour. The country was menaced with attack from the combined navies of Europe, amounting in all to 180 sail of the line, and of that immense fleet they were well aware the Baltic squadron would form the right wing. No time was to be lost; in a few days an overwhelming force would to all appearance be assembled on the shores of the Great Belt, and if ferried over to Zealand might enable the Danish Government securely to comply with the requisition of the combined Emperors, and bid defiance to all the efforts of Great Britain. In these circumstances, on the 19th July, 1807, the Cabinet took the resolution of demanding the surrender of the Danish fleet, to be held in deposit in an English port as a pledge of neutrality, with the determination if this surrender were refused, to seize it forcibly, and convert what would otherwise be used against them into a valuable addition to their own navy.

Fortunately the troops which had been preparing to proceed to the Baltic were soon ready for embarkation. The first body of troops (mostly composed of Germans), as noted above, had already been despatched and had reached Stralsund and Rugen; and before the end of July, 20,000 men, including cavalry and a formidable train of artillery, were embarked or on the march to the sea coast. A large naval force was also assembled to act as occasion might require, and this was speedily added to with extraordinary expedition.

On the 26th July, Admiral Gambier, with the principal division of the fleet, consisting of seventeen ships of the line, with twenty-one frigates, sloop, bomb-vessels, and gun-brigs, set sail from Yarmouth roads. He came to anchor in the Sound on the 3rd August, having on his way, on the 1st August, detached Commodore Keats in command of several vessels to the Great Belt to prevent the passage of troops from the Continent to the island of Zealand. On the 4th August Admiral Gambier was joined by another fleet consisting of eight sail of the line, with frigates, sloops, and other light vessels. Previous to entering the Sound, it had been ascertained from the Commandant of Cronenburg that as he had received no instructions to oppose the passage of the fleet, he did not intend to do so.

The troops of the expedition from England left about the end

1807. Expedition to Copenhagen.

of July in three divisions: from the Downs, from Harwich, and from Hull. On the 9th August the three divisions were assembled in the Elsinore roadstead. Lieutenant-General Burrard had been hitherto the senior officer, but on the 12th August Lord Cathcart arrived from Rugen and took over the chief command. By the time the transports with the troops from Rugen had joined, the expedition consisted of 25 sail of the line and upwards of 40 frigates, sloops, bomb-vessels, and gun brigs; making a total of about 65 vessels of war, exclusive of 377 transports, measuring 78,420 tons, and conveying about 27,000 troops, a great part of whom were Germans in British pay.

The troops were divided into two Divisions and a reserve (of four regiments), and comprised, in addition the German legion, which latter, however, did not all arrive till the 21st August. The artillery consisted of ten companies of British gunners, and with the German legion there were two troops horse artillery and four companies of foot artillery. There were also three regiments of German cavalry. (For a more detailed statement of this force see Appendix A., Page 55.)

It was no part, however, of the design of British Government to precipitate hostilities; they wished to gain the object in view by diplomatic arrangements rather than by actual force. Mr. Jackson was, therefore, sent as envoy with the expedition, and he at once proceeded to Kiel and requested an audience of the Prince Regent of Denmark. He then made known the demands of the British Government, which were that the fleet should be deposited in pledge in an English port, under an obligation of restitution, at the conclusion of a general peace. These demands being indignantly refused by the Prince, the envoy had to declare that force would be employed. The Prince sent a messenger to Copenhagen carrying orders to have the city put in a state of defence, and these instructions were received on the evening of the 10th August. Early on the following morning work began, and at noon the Prince arrived. He, however, left again on the 12th, after entrusting the defence of the city to Major-General Peyman.

The regular force at this time in the city and suburbs of Copenhagen, of which the population in the preceding March was estimated at upwards of 100,000 souls, has been variously stated at from 3,000 to 10,000 men; but the account that appears to be the most worthy of credit places the number at 5,510, including an organised Militia force of 2,000 men. This was exclusive of sailors and of 3,000 armed citizens; so that the whole force, regular and irregular, amounted probably to 12,000 men. The main Danish army, of more than double that amount, was encamped in Holstein.

The sea defence of the port consisted of the Trekonen pile battery situated at the distance of 2,000 yards in a north-easterly direction from the harbour (which runs like a canal through the centre of the town), and mounting 68 guns, besides mortars; a pile battery in advance of the city mounting 36 guns and 9 mortars; the citadel itself mounting 20 guns and 3 or 4 mortars; and the holm or arsenal battery mounting 50 guns and 12 mortars; total, 174 guns and 25 mortars.

1807. Expedition to Copenhagen.

It has already been remarked that the harbour divides the city into two parts: the largest division is situated on the side of Zealand; the smaller, with the dockyard and arsenal, on that of Amack. The works defending the former part reached from sea to sea, and were three miles in extent. The breadth of the ditch was from 40 to 60 yards, and in no part had it less than 6 feet of water. There were three entrance gates to the town on this side; in front of each was an advanced work with a double wet ditch; some of the curtains were defended by ravelins; the western flank of this portion rested on the sea which separates the islands of Zealand and Amack in that quarter; and the eastern flank was protected by the citadel above referred to, a very strong work with a double ditch; it was against this part of the town that the operations of the assailants were directed. The ground in front of the works, owing to the existence of a large lake, passable at a few points only, also greatly assisted the means of defence, but the place was in no way prepared to resist an attack: the ramparts were unarmed and the fleet unequipped.

Near the Trekonen battery, and in front of the harbour, the Danes had the blockship "Mars," of 64 guns, 4 20-gun praams, 2 floating batteries, and from 25 to 30 gunboats, each of the latter mounting 2 heavy guns. The fleet in the dockyard was not in a serviceable condition.

The negotiations with the Court of Denmark terminated on the 13th August. On the 14th the weather was unfavourable for taking up a position for disembarking troops, but on the 15th the men-of-war and transports moved up to the bay of Vedbeck, a village about midway between Elsinore and Copenhagen, and eight miles to the north of the latter place. Major-General Spencer's Brigade proceeded under convoy of a division of the fleet higher up the Sound to cause a diversion. By 5 A.M. on the 16th the reserve, with some artillery, was landed in the bay of Vedbeck, and the heights in front immediately occupied; the remainder of the infantry and some more artillery landed during the forenoon, and a squadron of Light Dragoons of the German legion closed the disembarkation. No opposition was made to the landing; a few patrols of Danish cavalry watched the proceedings, and retired on the approach of the advanced guard.

Lord Cathcart and Admiral Gambier issued a joint proclamation, explaining the objects of the expedition, and promising that public and private property should be respected, and that no harm should be done to individuals so long as the proceedings of the troops were not opposed, and provided the inhabitants remained quietly in their homes.

On the evening of the 16th the army commenced its advance in three columns towards Copenhagen; the reserve on the right by Hœrum to Lyngby; the centre, by Eremitage and Fortunen, to Jœgersborg; and the left by the sea coast to Charlottenlund. At night the troops took up, unmolested, a position as follows:—the right was at Lyngby, which was occupied by the reserve under Sir A. Wellesley, and from that place the troops extended through Jœgersborg to Charlottenlund; the line was about three miles long, the right resting on a lake, and the left on the sea at about four miles from Copenhagen. 1807. Expedition to Copenhagen.

On the morning of the 17th August the troops again advanced; the reserve, under Sir A. Wellesley, inclined to its left and halted in front of Jœgersborg; Sir G. Ludlow's Division inclining at the same time to the right crossed the head of the reserve, and proceeded by Gladsaxe and Vanloes, to the road leading from Roeskilde to the capital. As soon as the head of the column entered this road a detachment was sent to hold a strong fort in the rear at the Dam House, which commanded the approach to Copenhagen. Sir G. Ludlow's Division was established at Fredericksburg; the left, under Sir D. Baird, rested on the sea; and the reserve was placed in rear of the centre; Copenhagen was thus completely invested. Two brigades of the German legion remained at Charlottenlund to cover the disembarkation of the cavalry and siege artillery, as well as to secure the rear of the left; whilst the rear of the right was protected by the 28th Regiment, which took up a position behind Fredericksburg. General Spencer's brigade landed the same day at Skorreshard. Headquarters were established at Hellerup, in rear of the centre, and thus all the arrangements preparatory to a siege were made without any molestation on the part of the Danes.

About noon on the 17th, however, they made an attempt to dislodge our pickets advanced towards the town on the left, on which occasion their gunboats acted in co-operation with the troops. This attempt was easily frustrated by General Spencer's pickets, and by the troops advanced to their support. On the following day some fighting took place between the Danish and English gun-vessels, but with little effect on either side.

On the 18th the whole of the cavalry, being disembarked at Skorreshard, moved to Charlottenlund, Jœgersborg, and Vanloes, pushing their pickets into the country, and establishing a chain of posts well to their front, supported by a battalion of the German legion under the command of Brigadier-General von Decken.

On this day a summons was sent to General Peyman, but it produced no effect.

On the 19th the enemy's gunboats again attacked the English left, but were driven off by the field pieces; and the frigates and gun-brigs of the fleet, seizing the advantage of a favourable breeze, took their station near the entrance of the harbour, within range of the town.

The works were now carried on with great vigour, especially on the left of the attack. Brigadier-General von Decken surprised the post of Frederickswork, where 860 officers and men surrendered and were liberated on condition of not serving again during the war. A considerable number of small arms were found there, and their capture prevented the arming of the peasantry.

On the 21st August Lieutenant-General the Earl of Rosslyn, who had arrived from the Isle of Rugen with several battalions and two regiments of Light Dragoons of the German legion, disembarked in the north of Kioge Bay, a few miles to the south-west of Copenhagen. The disembarkation was covered by five companies of the 28th Regiment, and a detachment of cavalry. The cavalry patrolled

1807. Expedition to Copenhagen.

the country in the neighbourhood of Roeskilde, and thence towards the sea; the infantry remaining at the village of Vallensbek, within two miles of the coast to act as occasion required. Although a force of about 5,000 Danes, under General Cartinskiold, was assembled near the town of Kioge, no interference was attempted with Lord Rosslyn's disembarkation.

Great progress was now made in the works on the English left; new batteries were erected, and battering train and siege stores landed. Admiral Gambier declared the island of Zealand, and the other islands contiguous thereto, in a state of blockade. A large body of troops had assembled in Fyen, which is the largest of the Danish islands next to Zealand, from which it is separated by the Great Belt, about twelve miles across at the narrowest part. Several attempts were made to transport troops from this island, as well as from the Continent, to reinforce the Danes in Zealand; but they were all prevented by the vigilance of Commodore Keats.

On the 22nd August the army was reinforced by a Brigade, consisting of the 7th and 8th Regiments of infantry under Brigadier-General Macfarlane, which had landed the preceding evening at Skorreshard; they encamped in the rear of headquarters.

On the following day the Danish gunboats and praams attacked the advanced squadron of gun-brigs and bomb-vessels, which had taken a position near the entrance of the harbour, and after a combat of several hours forced them to retire. The same day Lord Rosslyn's force joined the army and took up its position in second line, covering the centre.

On the 24th, at daybreak, the army was under arms for the purpose of taking up a position closer to the town; the centre advanced to the high ground near the road, which runs in a direction parallel to the defences of Copenhagen, to Fredericksberg, occupying that road and the posts beyond it. The Guards at the same time occupied the suburbs between Fredericksberg and Copenhagen, flanked by a detachment of the 79th Regiment, and suppooted by one wing of the 28th Regiment in the village of Valdby. All the pickets of the enemy fell back to the lakes or inundations in front of the place, the English pickets occupying the position they abandoned.

In the afternoon the garrison showed itself in all the avenues leading from the town, apparently with the design of recovering their ground, or of burning the suburbs. The several Generals immediately drove them in, each in his own front, and at the same time seized all the suburbs on the north bank of the lakes, some of which were within 400 yards of the place. During these operations Sir D. Baird's division turned and carried a redoubt which the enemy had been some days constructing, and which was the same night converted into one of the besiegers' works. In the evening the Danes set fire to the suburbs leading to the western gate.

In consequence of the general success attending this day's movements, the works which had been planned and begun by the assailants were abandoned, and a new line taken up within about 800 yards of the place, and nearer to it on the flanks.

On the 26th, and following day, some fighting took place

# EXPEDITION TO

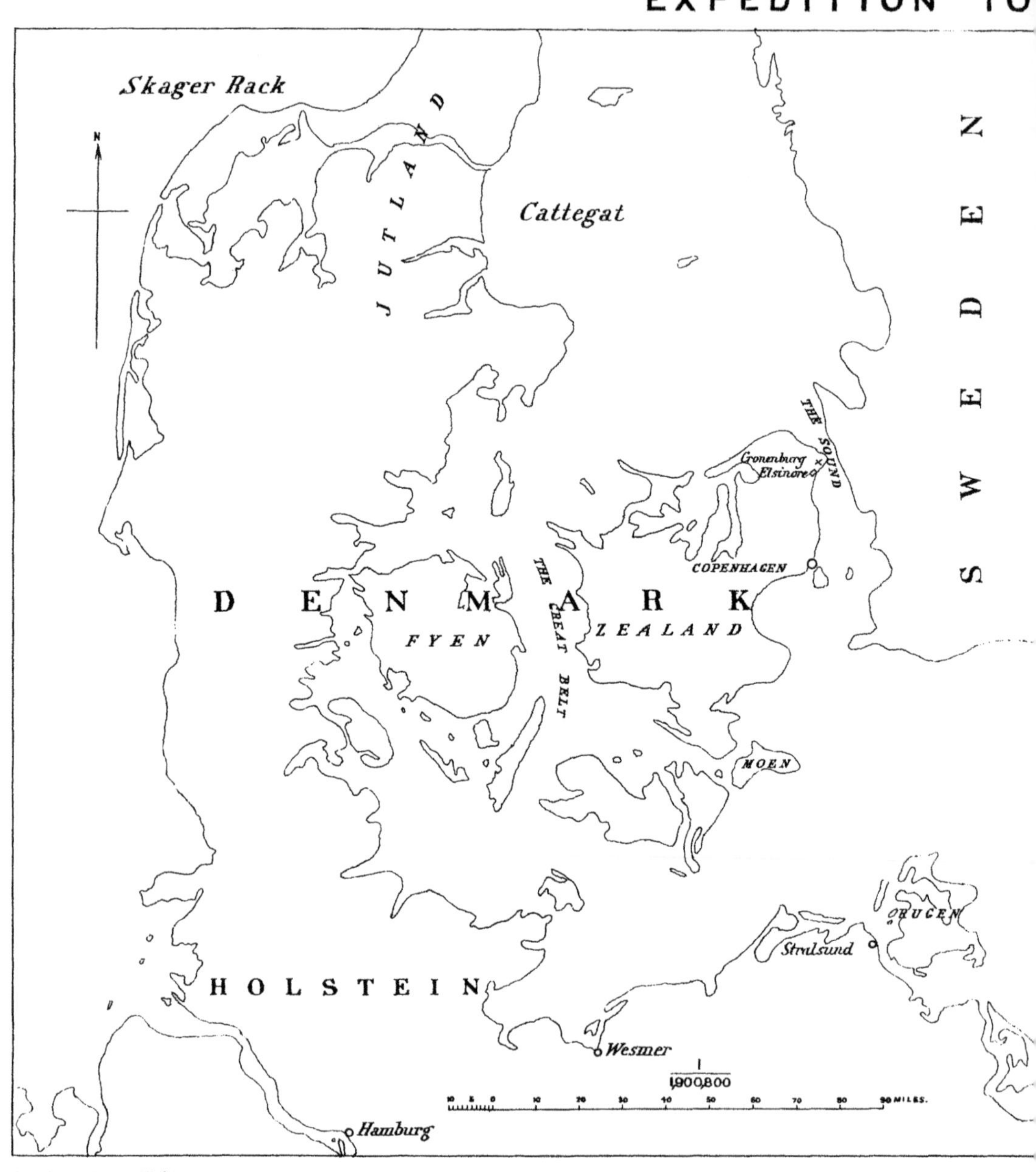


Intelligence Branch No 308.

# OPENHAGEN 1807.

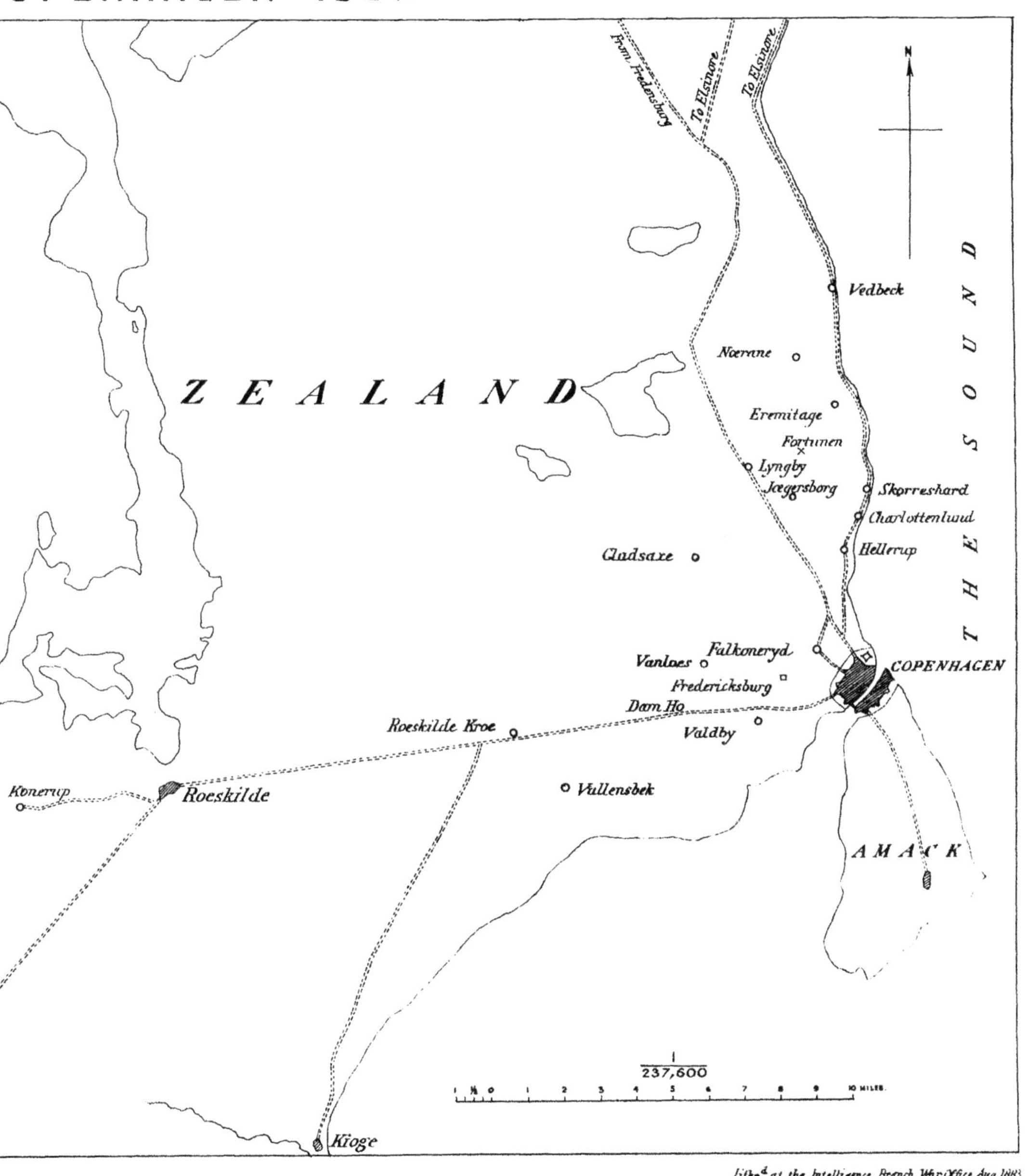


Lithod at the Intelligence Branch War Office Aug 1883

1807. Expedition to Copenhagen.

between the Danish gunboats and the troops on the right flank of the attack; but the gunboats were eventually driven off with loss by our artillery.

On the 26th August Sir A. Wellesley, with the reserve, eight squadrons of cavalry, and the horse artillery under Major-General Linsingen, a battalion of the German legion, and some artillery, the whole force being about 5,000 men, marched to Roeskilde Kroe to observe the force assembled under General Castenkiold, and to prevent its making any attempt against the army before the town.

On the 27th, Sir A. Wellesley having formed his force into two divisions, the one commanded by himself, and the other by Major-General Linsingen, marched from Roeskilde Kroe to attack the enemy at Konerup. It was found, however, that the enemy had moved off by his right to Kioge, and consequently Sir A. Wellesley fell back to Roeskilde Kroe, establishing his right there and extending his left towards the sea, so as to cover the right of the besieging army; Major-General Linsingen took post with his division at Roeskilde.

Early on the 29th the troops under Sir A. Wellesley moved against the enemy at Kioge. It had been arranged that Major-General Linsingen was to turn the enemy's left, whilst Sir A. Wellesley advanced along the sea road to attack the front of their position. The Danes were found in position to the north of Kioge, but they did not long stand the combined attack made against them. They fled in confusion, and with great loss; about 1,600 were killed and wounded, and 60 officers and 1,500 men taken prisoners. It must, however, be recollected that the force of from 6,000 to 7,000 men they had collected consisted principally of armed peasants, and thus although the victory gained over them was therefore not very glorious, it was still of considerable importance, as it rendered the rear of the besieging army secure from attack.

The officers who were taken prisoners were liberated on parole, and the men were distributed amongst the ships-of-war, with a view to deter the peasants in the country from rising in arms.

General Castenkiold, after his defeat, retired with the remains of his army to the small island of Moen, lying to the south of Zealand. Sir A. Wellesley moved into the interior to disarm and quiet the country.

Owing to the advanced position of the attack, St. John's Hospital, which was full of patients, became exposed to the fire from the outposts of both armies. General Peyman asked for an armistice of thirty-six hours to remove the sick into the town, and was offered four hours for this purpose by Lord Cathcart. The offer was refused by General Peyman. On the 29th a truce was agreed on to remove the Danish sick to the English lines; they were at first removed to a chapel, and afterwards sent into the country.

On the 30th August the batteries were nearly finished, the platforms laid, and two-thirds of the ordnance mounted; sailors from the fleet assisted in the work, and guns from the ships were landed to be placed in the batteries. On the 31st all the batteries, with one exception, were armed and completed.

1807. Expedition to Copenhagen.

Alarmed at the progress of our works, General Peyman made a sortie on the 31st August. The force consisted of three battalions of infantry, the Chasseurs of the King's Guard, the riflemen of other regiments, and the huntsmen from the country; in all between 2,000 and 3,000 men, supported by eight guns. They moved out from their right to reconnoitre a garden in front of the citadel, and to destroy the batteries, which it was supposed we had erected there. The advanced picket gallantly opposed their progress until supported by the other pickets of Sir D. Baird's division. This General officer conducted the affair in person, and immediately drove in the enemy, who retired in confusion through the gates of the citadel. Our loss on this occasion was trifling, but the Danes lost nearly 100 killed and wounded, General Peyman being among the latter.

On the 1st September the besiegers' works, including the mortar batteries, were ready for action. Though the Danes had kept up a fire from the walls and outworks with cannon and musketry, and had inflicted some trifling losses on the assailants, they had not succeeded in retarding the progress of the works.

On this day the General Commanding and Admiral Gambier addressed a joint summons to General Peyman, and offered the same terms as had been at first proposed. These terms were, however, rejected by General Peyman, who declared his intention of defending the town to the last extremity. His answer was not received until late at night, and as communication had then to be made with the Admiral on board the Flagship, the correspondence was not closed until noon on the following day.

About half-past seven in the evening of the 2nd September, the English batteries, consisting of forty-eight mortars and howitzers, and twenty 24-pounders, opened against the town, which was immediately set on fire in different places; the navy also shelled the town, and the firing on shore continued without interruption till 8 A.M. on the 3rd. The enemy's fire was very slack during the night. The fires in the town were soon extinguished, the engines being in good order, and in charge of 5,000 well-drilled firemen. The bombardment at first told more on the inhabitants, numbers of whom were killed, than on the houses. During the 3rd all the batteries kept up a shell fire at intervals, from the effects of which the people fled in large numbers to the island of Amack. That small island received 32,000 fugitives from the capital, and as these consisted of women, children, and aged people, they were allowed to remain there in safety, no attempt being made to bombard that island.

On the night of the 4th our batteries continued the bombardment with renewed vigour; a large timber yard was set on fire by red-hot shot, and the Fruekirke, the handsomest church in the town, took fire. The fire engines had now become useless from frequent use, or from the effect of shell fire; the firemen were killed or wounded, or worn out with fatigue, and the city was at length abandoned to the flames. Very few shot or shell were fired from the place, and the troops appeared to have deserted their works. Towards daybreak on the 5th one-third of the town appeared in a blaze, and the Fruekirke fell in one immense mass to the ground. Several other public buildings were also destroyed, and numerous private houses burnt to the ground.

1807. Expedition to Copenhagen.

On the evening of the 5th, after the bombardment had lasted three nights and three days, the Governor sent out to propose an armistice for twenty-four hours, to settle the preliminaries of a capitulation. Lord Cathcart was unwilling to accede to an armistice, but he despatched Lieutenant-Colonel Murray into the town to receive any proposal the Governor was inclined to make, upon which articles of capitulation might be founded. On the following morning, the Governor having accepted the proposal of delivering up the Danish fleet as the basis of negotiation, an armistice was declared until the negotiations were concluded.

Preliminaries being thus far arranged, Major-General Sir A. Wellesley, Sir Home Popham, and Lieutenant-Colonel Murray were appointed to prepare and sign articles of capitulation. The capitulation was drawn up in the night of the 6th–7th September, and the ratification was exchanged in the course of the morning.

The articles of capitulation, nine in number, were very favourable to the Danes. (For their full text see Appendix B, Page 56.)

It was stipulated that the British forces should be put in possession of the citadel and dockyard; that all ships-of-war, with the naval stores belonging to the King of Denmark, should be delivered into the charge of such persons as should be appointed by the British Commanders-in-Chief, and that the same persons should be put in immediate possession of the dockyards and all the buildings and storehouses belonging thereto; that the British storeships and transports should be allowed to go into the harbour for the purpose of embarking stores and troops; that as soon as the Danish ships, &c., should be removed from the dockyard, or within six weeks from the date of the capitulation, the British troops should deliver up the citadel in the same state in which it was found; and that within the same time, or sooner if possible, our army should embark from the island of Zealand; that all prisoners on both sides should be restored; that all officers on parole should be released unconditionally, and that all English property that had been seized should be restored to its owners.

The total loss of the British troops during this expedition was 4 officers and 38 non-commissioned officers and men killed, 6 officers, 139 non-commissioned officers and men wounded, and 24 non-commissioned officers and men missing, making, with the loss of the British afloat, 56 killed, 179 wounded, and 25 missing.

The loss on the part of the Danes on board the gun-vessels, and in the different skirmishes outside the city, appears by their own accounts to have been about 250 killed and wounded, and a great number besides were taken prisoners. Their loss within the city, including men, women, and children, was about 1,500. The number of houses wholly destroyed was officially stated at 305, but scarcely a house escaped from the effects of the bombardment.

1807. Expedition to Copenhagen.

The British troops took possession of the citadel, dockyard, and arsenal on the day the capitulation was signed.

The Danish ships were fitted out with great expedition and in nine days fourteen sail of the line were towed from the harbour to the roadstead. In the space of six weeks the three remaining ships of the line, with the frigates and sloops, were removed to the roadstead, and the arsenal and its storehouses cleared of masts, spars, timber, and other naval materials.

On the 20th October, by which time all the ships and small craft were out of Copenhagen harbour, the last troops of the British army re-embarked with the utmost quietness and without a casualty; and on the 21st, in the morning, the British fleet sailed from Copenhagan Roads in three divisions. It had been arranged that in going down the Sound the castle of Cronenburg should abstain from hostilities, and allow the fleet, which kept as much as possible on the Swedish side of the Channel, to pass in safety. On entering the Cattegat, the weather became boisterous, and led to the destruction of all the Danish gunboats but three. After this, the fleet proceeded without further accident, and at the close of the month reached Yarmouth and the Downs. A large proportion of the seamen of the fleet having been distributed amongst the Danish ships, the assistance of the troops was necessary to navigate the fleet to the Downs and to Portsmouth, and this circumstance prevented any part of the army being landed at Yarmouth. Owing to unfavourable weather after reaching the English coast, the disembarkation of the troops was not completed until the middle of November.

The Copenhagen expedition caused great excitement throughout Europe; it was at the time universally condemned as an unjustifiable violation of the law of nations. Public opinion was also much divided on the subject in England, both as to the lawfulness of the expedition, and the justice of retaining the prizes which had been made. The non-production of the secret articles of the Treaty of Tilsit, which the ministers stated were in their possession, caused the public for a long time to doubt the necessity for the vigorous measures taken by the ministry, and in this respect full justice was not done to their policy until the year 1817, when, on the death of the persons who had given the information, the secret article was prodnced.

The expediency and necessity of the strong steps taken by us against Denmark, as a measure of self-preservation, have since been fully acknowledged. Writers on the Law of Nations are clear that in such circumstances as the Danish fleet was placed, its seizure was perfectly justifiable. "I may," says Grotius, "without considering whether it is merited or not, take possession of that which belongs to another, if I have any reason to fear any evil from his holding it; but I cannot make myself master or proprietor of it, the property having nothing to do with the end which I propose. I can only keep possession of the thing seized till my safety is sufficiently provided for."

This was precisely what the English Government had proposed to Denmark.

## APPENDIX A.

1807. Expedition to Copenhagen. App. A. Composition of Expeditionary Force.

Lieutenant-General Lord Cathcart, Commander-in-Chief.
,, Burrard, Second in Command.

STAFF.

Colonel Hope, Deputy Adjutant-General.
Lieutenant-Colonel Murray, Deputy Quartermaster-General.
Major Macdonald, Military Secretary.

RIGHT DIVISION.

Lieutenant-General Sir G. Ludlow.
Major-General Finch.
One battalion Coldstreams; one battalion 3rd Guards.
Brigadier-General Warde.
1st battalion 28th Regiment; 79th Regiment.

LEFT DIVISION.

Lieutenant-General Sir D. Baird.
Major-General Grosvenor.
1st battalion 4th Regiment; 1st battalion 23rd Regiment.
Major-General Spencer.
32nd Regiment; 50th Regiment; 82nd Regiment.
Brigadier-General Macfarlane.
7th Regiment; 8th Regiment.

RESERVE.

Major-General Sir A. Wellesley.
Acting Brigadier-General Colonel Stuart.
1st battalion 43rd Regiment; 2nd battalion 52nd Regiment;
1st ,, 92nd ,, ; 1st ,, 95th ,,

ROYAL ARTILLERY.

Major-General Bloomfield.
10 companies.

ROYAL ENGINEERS.

Colonel D'Arcey.

GERMAN LEGION.

Lieutenant-General the Earl of Rosslyn.
Major-General Linsingen.
,, Dreskeil.
Brigadier-General von Decken.
3 regiments Cavalry; 10 battalions Infantry; 2 troops Horse Artillery; 4 companies Foot Artillery.

The combined British and German force amounted to 27,000 men.

1807. Expedition to Copenhagen. App. B. Articles of Capitulation.

## APPENDIX B.

Articles of Capitulation for the Town and Citadel of Copenhagen, agreed upon between Major-General the Right Honourable Sir Arthur Wellesley, K.B.; Sir Home Popham, Knight of Malta, and Captain of the Fleet; and Lieutenant-Colonel George Murray, Deputy Quartermaster-General of the British Forces; being thereto duly authorised by James Gambier, Esq., Admiral of the Blue, and Commander-in-Chief of His Britannic Majesty's Ships and Vessels in the Baltic Sea, and by Lieutenant-General the Right Honourable Lord Cathcart, K.T., Commander-in-Chief of His Britannic Majesty's Forces in Zealand and the North of the Continent of Europe, on the one part;—and by Major-General Walterstorff, Knight of the Order of Dannebroge, Chamberlain to the King, and Colonel of the North Zealand Regiment of Infantry; Rear-Admiral Luthen, and J. H. Kerchhoff, Aide-de-Camp to His Danish Majesty; being duly authorised by his Excellency Major-General Peyman, Knight of the Order of Dannebroge, and Commander-in-Chief of His Danish Majesty's Forces in the Island of Zealand, on the other part.

Article I. When the capitulation shall have been signed and ratified, the troops of His Britannic Majesty are to be put in possession of the citadel.

Article II. A guard of His Britannic Majesty's troops shall likewise be placed in the dockyards.

Article III. The ships and vessels of war of every description, with all the naval stores belonging to His Danish Majesty, shall be delivered into the charge of such persons as shall be appointed by the Commanders-in-Chief of His Britannic Majesty's Forces; and they are to be put in immediate possession of the dockyards, and all the buildings and storehouses belonging thereto.

Article IV. The store-ships and transports in the service of His Britannic Majesty are to be allowed to come into the harbour for the purpose of embarking such stores and troops as they have brought into this island.

Article V. As soon as the ships shall have been removed from the dockyard, or within six weeks from the date of this capitulation, or sooner, if possible, the troops of His Britannic Majesty shall deliver up the citadel to the troops of His Danish Majesty, in the state in which it shall be found when they occupy it. His Britannic Majesty's troops shall likewise, within the before-mentioned time, or sooner if possible, be embarked from the Island of Zealand.

Article VI. From the date of this capitulation hostilities shall cease throughout the Island of Zealand.

Article VII. No person whatsoever shall be molested; and all property, public or private, with the exception of the ships and vessels of war, and the naval stores before mentioned, belonging to His Danish Majesty, shall be respected; and all civil and military officers in the service of His Danish Majesty shall continue in full exercise of their authority throughout the Island of Zealand; and everything shall be done which can tend to produce union and harmony between the two nations.

Article VIII. All prisoners taken on both sides shall be unconditionally restored, and those officers who are prisoners on parole shall be released from its effect.

ARTICLE IX. Any English property that may have been sequestered in consequence of the existing hostilities, shall be restored to the owners.

1807. Expedition to Copenhagen. App. B. Articles of Capitulation.

This capitulation shall be ratified by the respective Commanders-in-Chief, and the ratifications shall be exchanged before twelve o'clock at noon.

Done at Copenhagen, this 7th day of September, 1807.

(Signed) ARTHUR WELLESLEY.
HOME POPHAM.
GEORGE MURRAY.

Ratifié par moi (Signé) PEYMAN.

---

## EXPEDITION TO THE SCHELDT (WALCHEREN), 1809.

1809. Walcheren Expedition.

IN 1809, Napoleon was making every effort to establish at Antwerp a commercial rival to London, and to avail himself of the naval resources which the possession of Holland and the Scheldt placed at his command, thus keeping the "pistol presented at the breast of England," ready charged for any future contingencies.

The English Government thought, by striking a blow in the Scheldt, to frustrate Napoleon's hopes of maritime rivalry at the outset, and further, to detain in Holland a portion of the forces the Emperor was at that time setting in motion against Austria.

To effect these objects the English Government, in the summer of 1809, determined to send an expedition to the Scheldt, and on the 16th July Lieutenant-General the Earl of Chatham, K.G., was appointed to the command of a land force destined to attack and destroy the naval forces and establishments at Flushing, Antwerp, and Terneuse, and in the island of Walcheren, and to render the river unnavigable for ships-of-war.

The General's Staff was composed of the following officers:—

Lieutenant-General Sir Eyre Coote, Second in Command.
Colonel Carey, Military Secretary.
Colonel Long, Adjutant-General.
Lieutenant-General Brownrigg, Quartermaster-General.
Brigadier-General McLeod, commanding Royal Artillery.
Colonel Tyres, commanding Royal Engineers.
Mr. Robinson, Commissary-General.
Mr. Webb, Inspector-General of Hospitals.

The troops ordered to embark numbered some 39,000 men (rank and file only).*

Rear-Admiral Sir Richard Strachan was appointed to the com-

* First Division { Lieutenant-General Sir John Cradock. Major-General Graham.
Second " Lieutenant-General the Marquis of Huntly.
Third " Lieutenant-General Grosvenor.
Fourth " Lieutenant-General Fraser.
Fifth " Lieutenant-General Lord Paget.
Light troops, Lieutenant-General the Earl of Rosslyn.
Reserve, Lieutenant-General Sir John Hope.

1809. Walcheren Expedition.

mand of the fleet destined to convey the troops, and to co-operate with Lord Chatham's forces in the destruction of the naval establishments on the Scheldt.

The line-of-battle ships were not to go up the Scheldt, though the crews might be employed in manning small boats and ships of light draught.

A number of lighters, cutters, long boats, &c., from the dockyards, were to attend the expedition, and help in ascertaining and buoying the channel up the river.

Fifteen thousand troops from Portsmouth were to be carried in ships-of-war, and were destined for the attack on Flushing.

During the winter and spring the Admiralty had been busy collecting information of the enemy's movements in the Low Countries, and a *précis* of this had been furnished to the Government. Naval directions for the approach to Walcheren and the mouths of the Scheldt had been drawn up by Captain Bolton, R.N.

At this time Walcheren was held by the Dutch, whose troops, under General Bruce, garrisoned Middleburg and some forts on the northern shore. The fortress of Flushing, ceded to France in 1807, was occupied by a French garrison of 3,000 men under General Monnet. Of these, however, only a few hundred were French, the others belonging to foreign corps, viz: 1st battalion Irish, 1st Colonial battalion, 2nd battalion Prussians.

Cadzand, which formed part of the 24th Military District, commanded by General Chambarthac, was held by a few hundred of the National Guard, placed in coast batteries under General Rousseau, the two infantry battalions being quartered in Ghent to avoid the unhealthy climate.

The French ships in the Scheldt were commanded by Admiral Missiessy, and consisted of ten ships-of-the-line belonging to the Antwerp Fleet.

Had the English landed immediately at Blankenberg, and pushed along the high road from Bruges by Eecloo and Sas de Gand to Antwerp, they might in seventy-two hours have appeared before the Tête de Flandre and Fort Liefkenshoek. Had Liefkenshoek been carried by a *coup de main*, or had even batteries been thrown up on the banks of the Scheldt at its narrow part, the French Fleet would have been prevented from escaping up the river to Antwerp.

A division of 10,000 or 12,000 men sent from Blankenberg to Courtrai would have guarded the main road to Menin, and given timely notice of any approach of the enemy from France. Had the enemy then advanced in force, the English could have easily embarked in the numerous channels between Liefkenshoek and Terneuse under cover of their fleet. Instead of this bold action Lord Chatham followed the more cautious one of besieging Flushing, and by its possession securing a base in Walcheren for further operations along the north bank of the Scheldt.

Lord Chatham, with Sir Richard Strachan, sailed from the Downs on the 28th July, and anchored the same evening in the East Capelle Roads.

The want of pilots was being severely felt; Captain Owen, R.N., writing from the Downs on the 28th July, 1809, stated that being personally unacquainted with the navigation of the North Sea he must leave several ships behind. **1809. Walcheren Expedition.**

The ships carrying Lieutenant-General Sir John Hope's Division (the reserve) joined the following day; but a fresh gale from the westward having sprung up the fleet took shelter in the Roompot.

The left wing, under Lieutenant-General Sir Eyre Coote, which was destined for service in Walcheren, arrived on the 29th and 30th July.

A landing on the west coast of Walcheren with the view of forcing the passage of the Scheldt is only possible when the wind is moderate and from the east. At the time this expedition arrived, a fresh gale was blowing which raised such a surf that it was decided to abandon all idea of landing between Domburg and Zoutelande; and till Flushing was taken it was not thought prudent to send the fleet into the Scheldt. To the north-east of Walcheren there was sheltered water in the Veersche Gat, and though hitherto it had not been considered practicable for large ships, the Admiral decided to make the attempt. This was successfully accomplished, and the 1st, 4th, and 5th Divisions, under Sir Eyre Coote, landed without opposition on the Bree-Zand, about one mile to the west of the Fort-den-Haake, and occupied a position on the sandhills facing Oostkapelle. The 4th Division (General Fraser's) was detached to the left against Fort-den-Haake and the town of Veere. The fort was evacuated on the approach of the English; but Veere, which was strongly fortified and occupied by a garrison of 600 Dutch, held out till the morning of the 1st August, notwithstanding the fire from the gun-vessels.

During the disembarkation the navy rendered very great assistance by landing a party of officers and seamen, who dragged the guns over the sand, the horses not having been disembarked owing to the strength of the tide.*

The reserve under Sir John Hope remained on board ship in the Veersche Gat.

Commodore Owen's squadron, with the 2nd Division (the Marquis of Huntly's), had anchored in the Wielingen on the afternoon of the 29th July.

On the first approach of the English, General Rousseau, whose headquarters were at Breskens, was informed of their presence, and sent orders to bring up his two battalions by forced marches from Ghent to Groede. The chief defence of the coast of Cadzand was a battery near Breskens mounting some twenty guns and six mortars; besides this there were several unfinished and badly armed batteries. The old fort of Ysendick, which was well placed for a

* The following description of the landing-place at Bree-Zand may be of interest:—"To the north of the island (Walcheren), in front of some dunes of slight elevation, is the Bree-Zand, which offers the greatest facilities for landing. Its form is that of a segment of a circle, which has sides 300m. (328 yards), and an arc 200m. (218 yards). Everywhere a short distance from the shore, even at low tide, there are three, six, and eight fathoms water, so that frigates and brigs, by placing themselves at the extremities of the chord, flank with their fire a space where 6,000 men can be drawn up. The nearness of the dunes which join the beach besides allow the disembarking force to turn the right and left of any troops drawn up for the defence of the coast."

1809. Walcheren Expedition.

defence of Cadzand in case of a landing being effected, had been dismantled in 1804.

General Rousseau made the greatest show he could with his 300 National Guards, and during the night, placing his best men in the Breskens battery, took up a position with the remainder near Groede.

The tide would have enabled the English to land at 3 A.M. on the 30th July; but either the Commodore and Lord Huntly had been deceived as to the strength of the French in Cadzand, or else there were naval reasons for not risking the disembarkation; whatever may have been the reason, no attempt at landing was made.

In his defence before the Court of Inquiry, Commodore Owen gave as his chief excuse for not making a landing at Cadzand, the necessity for collecting sufficient boats to land 3,000 men at a trip.

By noon on the 30th the two battalions of the 48th provisional regiment and a battalion of the 65th reached Groede, having marched without a halt some forty miles.

These troops were seen from the tops of the ships of the English squadron in the Weilingen.

As soon as General Monnet heard of the approach of the English he ordered Brigadier-General Osten with three battalions and four light field guns (two 3-pounders, and two 64-pounders), to proceed directly to Oostkapelle in order to watch their movements, and if possible hinder their disembarkation.

There was not any field artillery available at Flushing, and Osten's guns were old pieces taken from the walls and drawn by farm horses.

During the 30th, General Bruce, who commanded in Walcheren for the King of Holland, abandoned the island and drew off the greater part of his troops to Fort Bath, at the eastern extremity of South Beveland.

On the 31st the English advanced towards Middleburg. Brigadier-General Osten opposed their march, but he could do little, and eventually drew back and re-entered Flushing in the evening.

The 1st and 5th English Divisions, under Major-General Graham and Lieutenant-General Lord Paget, that night occupied Meliskirke on the right, Crupskirke in the centre, and on the left St. Lauren.

General Fraser's Division (4th) remained in front of Veere.

A deputation arrived at the English headquarters during the day, stating that the garrison had been withdrawn, and offering terms of capitulation which were readily agreed to.

The French squadron under Admiral Missiessy had been ordered to take refuge in Flushing, but he wisely preferred entering the Scheldt. He remained off Flushing till the 31st, when he began to fear that the English might intercept his retreat by coming down the Sloe Passage, and as the wind was fair, gave orders to make sail up the river, and in the evening was above Fort Bath.

News of the English descent on the Scheldt reached Paris on the 31st, and great fears were entertained for Antwerp.

The whole of the northern provinces had been drained of troops

1809. Walcheren Expedition.

for the foreign wars; at this time there were no less than 700,000 soldiers under arms out of France, viz.: 300,000 in Spain, 300,000 in Germany, and 100,000 in Italy.

In addition to the troops already mentioned as forming General Rousseau's command, and the garrison of Antwerp, the 6th and 7th demi-brigades were quartered between Brussels and Boulogne and four battalions of different regiments at Louvain; the battalion of the Vistula and some squadrons of Polish Lancers, and several batteries under orders for the Danube, were also in the north, and forming part of Rampon's command, which extended from Picardy to Holland. The King of Holland, as constable of Antwerp, and in virtue of a commission given to him in 1806, was in nominal charge of that fortress, which at that time was only defended by the old works; the guns were not even mounted on the ramparts, and the garrison only numbered some 2,000 men.

Owing to the national spirit of economy both the Dutch fleet and army were in a very inefficient state. There were in Holland only 5,000 regular troops, the remainder being abroad—four regiments in Germany and two in Spain.

Neither Dutch nor Belgians were particularly well affected towards the French, the Berlin decrees having crippled their commerce and made them anxious for friendly relations with Great Britain. The first real victory of the English would have made the people of the Low Countries their allies; and the clergy, with the exception of the Bishop of Malines (Napoleon's nominee), were anxious for the success of the English.

The French Government, however, in the absence of Napoleon, took active measures for the defence of Antwerp.

Fouché was most energetic, and wished to call out immediately 100,000 National Guards and give the command to Bernadotte; but Clarke, the Minister of War, was doubtful whether the Emperor would approve of either calling ont the National Guard in such numbers, or giving so important a command to Bernadotte. Clarke maintained that the troops already in the Northern Departments would, with the aid of the National Guard d'Élite, which was already organised, be sufficient for the immediate defence, and he gave orders that all troops in the north should be directed on Cadzand and Antwerp, whilst from Paris he sent the 2nd demi-brigade of four battalions with the utmost haste on the 3rd and 4th August to Antwerp. General Moncey was to collect the Gendarmerie-à-Cheval of the northern provinces, and every man that could be spared from the camp at Boulogne was to march towards the threatened coast. By this means the Minister of War hoped to collect some 20,000 men; *i.e.*, 10,000 regular infantry, 5,000 National Guard d'Élite, 5,000 Gendarmerie and local artillery of this country.

King Louis was hastening to Antwerp, and when he reached that city Clarke hoped to have 30,000 men, including those at Boulogne, available for service on the banks of the Scheldt. Decaux was sent as Commanding Engineer, with instructions to put Antwerp in an immediate state of defence.

On the 1st August the English in Walcheren advanced against

1809. Walcheren Expedition.

Flushing in order to invest the town. The enemy made a firm stand, and were reinforced during the afternoon by a battalion of the 65th Regiment from Cadzand, but were finally driven within their lines. Major-General Graham carried the batteries of Dykeshook, Viggeter, and the Nolle; and Brigadier-General Houstoun drove back the enemy with the loss of four guns on the Middleburg road; Lord Paget's division occupied West Zoutberg.

During the action the light troops under Brigadier-General Baron Rotenburg, and the 3rd battalion Royals, and flank companies 5th Regiment, particularly distinguished themselves.

Veere surrendered in the morning, and General Fraser's Division was marched in the evening towards Ritthem, detaching troops to reduce Fort Rammekens.

According to Lord Chatham's despatch of the 2nd August, 1809, written from Middleburg, Commodore Owen's squadron, with Lord Huntly's Division, was still at anchor in the Wielingen Roads; but according to the French account it had already sailed for the Roompot. The latter is most probably the case, as General Rousseau could scarcely have passed over the battalion of the 65th from Breskens to Flushing had the English been threatening Cadzand.

Lord Rosslyn's Division (light), and General Grosvenor's (the 3rd), remained on board ship in the Veerschegat.

The reserve under Sir John Hope was conveyed some distance up the East Scheldt in boats, owing to the difficulty of navigation for larger vessels, and landed early on the 1st August, on the north side of South Beveland, between Wemeldinge and Kattendyke, and pushing on a detachment towards Goes, that town the chief place in the island capitulated, the garrison retiring towards Fort Bath. That night the right of the division occupied Goes, Kapelle, and Biezelinge.

The English in Walcheren settled down before Flushing and Rammekens; Lieutenant-General Grosvenor's Division was landed, and the troops were employed in making batteries, bringing up artillery stores, &c.

In South Beveland, Sir John Hope pushed on towards Fort Bath and occupied Waarden on the night of the 2nd, and the following day the English were masters of the whole of South Beveland except Fort Bath, in which General Bruce and 600 Dutch troops had taken refuge.

On Admiral Missiessy's arrival before Bath on the evening of the 31st July he found that the works were in a very neglected condition. The fort mounted thirty guns on a level with the water, and whilst with a resolute garrison it would present a very serious obstacle to a naval advance up the Scheldt, it offered but a slight defence against a land attack. The Admiral therefore urged its immediate repair, and placed a frigate across the Bergen-op-Zoom channel as an additional protection.

On the 1st August the French ships of the line were withdrawn towards Forts Lillo and Liefkenshoek. In these forts were placed French garrisons, the works were vigorously repaired, and stockades were formed in the river as protection against fire-ships.

General Bruce did not hold out long in Fort Bath, but deserted

# MOUTHS OF THE SCHELDT

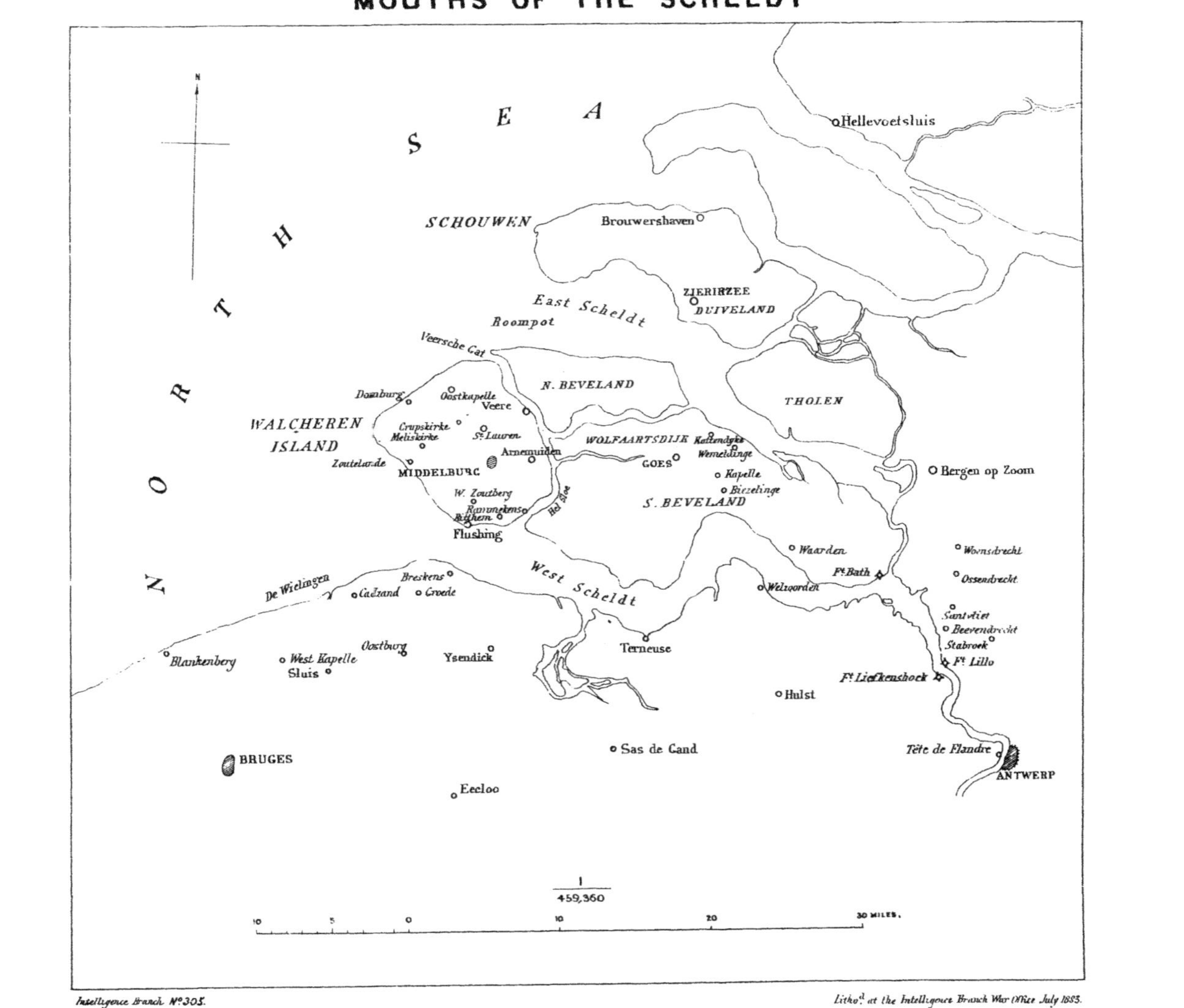


Intelligence Branch No. 305.

Litho.d at the Intelligence Branch War Office July 1883.

it without striking a blow; and according to the French accounts without even disabling the guns, which was done by a party of seamen landed by Admiral Missiessy, who entered the fort before the English were aware of its being deserted.

1809. Walcheren Expedition.

A patrol of thirty men sent by General Hope to reconnoitre the southern shore of the island, entered the fort on the 4th August.

General Bruce, who retreated to Bergen-op-Zoom, was refused admission into that town on account of his cowardice in deserting Fort Bath.

Whilst Sir John Hope was carrying on these movements in South Beveland, the preparations for bombardment of Flushing were being steadily pushed forward, though hindered by the heavy rain.

Fort Rammekens surrendered on the evening of the 3rd August, just as the batteries were ready to open fire.

The whole attention of the English in Walcheren was now centred on the reduction of Flushing, the guns of which prevented the English squadron entering the West Scheldt except by the slow and tortuous passage through the Sloe Channel.

Sir John Hope had sent earnest requests for a flotilla to guard Fort Bath and attack the French works beyond, but Lord Chatham was unable to comply with this request, and on the evening of the 5th, a French flotilla of twenty-eight gunboats descended the Scheldt and proceded to bombard Fort Bath.

The French had left twelve spiked guns in the fort, and John Skinner, a private in the 1st Guards, offered to unspike them with tools contrived by himself. Skinner* eventually accomplished his work under the enemy's fire, and the guns were immediately served and directed against the French gunboats, which after some time were forced to withdraw.

The French garrison in Flushing was very enterprising, and was continually receiving reinforcements from Cadzand, whence General Rousseau sent, on the 4th August, two battalions of the 8th provisional regiment; on the 5th, one battalion of the 48th regiment from Antwerp; and on the 6th, a detachment of 200 men from various corps.

On the 5th, about 3 o'clock in the afternoon, the garrison made a sortie in force against the right of the English line. The attack was repulsed by the troops of Lieutenant-General Graham's Division under Colonel Hay, but with loss of 14 killed, 141 wounded, and 5 missing.

Lord Chatham was anxious to send more troops into South Beveland, and ordered Lord Rosslyn's and Lord Huntly's Divisions, with the light brigades of artillery, to join the reserve under Sir John Hope. The cavalry and ordnance ships were to pass through the Sloe and work up the West Scheldt, and Lieutenant-General Grosvenor's Division was to follow as soon as it could be spared from Walcheren.

Lord Chatham excused the delay in the execution of the above

* On the return of the Guards to England, the Duke of York and the officers of the 1st Guards presented Private Skinner with a medal for his gallant conduct at the attack on Fort Bath.

**1809. Walcheren Expedition.**

orders alleging the want of a sufficient naval force to protect the transports in the Scheldt; and stated that this absence was due to difficulty of navigation and contrary winds.

The enemy endeavoured to check the besiegers' progress by opening the sluice gates; but those at Veere being in the hands of the English, the inundations were kept within bounds. The presence of the English army caused great distress to the inhabitants of Walcheren. There were very few cattle in the island, and the milch cows had to be killed to supply rations, which were only paid for at the market price of beef before the war. The farm carts, horses, and drivers had also been employed in the transport of stores, &c., when they were urgently wanted to gather in the harvest.

On the 9th August, Lord Rosslyn's and Lord Huntly's Divisions landed in South Beveland; and Lord Rosslyn assumed command of all troops in that island.

About this time intelligence was sent to Lord Chatham from London that the Duke of Brunswick was expected in Heligoland with a corps from the Weser, and that His Serene Highness had expressed a wish to serve on the Continent under Lord Chatham's orders. The Duke never joined the army on the Scheldt, as a few days later a letter was sent from Downing Street stating that his troops were *en route* for England.

It was considered at home that the possession of Walcheren would offer a good entrepôt for English merchandise, and that means of smuggling them into the Continent through Holland would be readily found. Several merchants had already applied for licenses to trade at Walcheren.

In writing to inform Lord Chatham of this fact the Board of Trade stated that whilst they had not any power to refuse the licenses, they had taken the opportunity of inserting a clause that the holders of licenses must conform to the regulations of the military and naval authorities actually commanding at Walcheren.

This decision is worth noting, as it shows that the civil power in England acknowledges the authority of military and naval commanders, during active operations in foreign countries, to regulate the trade and proceedings of English subjects who may be within the districts occupied by the British arms.

On the morning of the 13th the batteries were completed. The frigates and bomb-vessels moved in close to the town, and at 1.30 P.M. in the afternoon the bombardment of Flushing was begun by land and sea.

During the night an additional battery opened, and an advanced trench in front of General Graham's position was carried by a detachment of the 14th Regiment, and King's German Legion under Colonel Nicolls.

On the following morning the line-of-battle ships stood in and fired on the sea defences. After some hours the enemy's fire, which at first had been well sustained was noticed to slacken. A summons was sent to the Commandant Général de Division Monnet; but as no answer was returned, the batteries again opened fire.

1809. Walcheren Expedition.

In the evening, detachments of the 36th and 71st Regiments and the light battalion King's German Legion, under Colonel Pack, carried at the point of the bayonet an advanced battery in front of General Fraser's position.

The bombardment had fired the town in several places. Early in the morning, 14th August (2 A.M.), the French demanded a suspension of hostilities. During the day terms of capitulation were agreed upon between the English Commissioners, Captain Cockburn, R.N., and Colonel Long, Adjutant-General; and the French Commissioners, Captain Montonnet, of the Artillery, and Captain l'Evèque, of the Engineers. The ratifications were exchanged at midnight.

The French troops in Flushing were very indignant at the terms of capitulation, and were ready to renew hostilities; but any attempt was prevented by the early occupation of the gates the next morning by the English.

The garrison marched out as prisoners of war the following day, and were immediately embarked at Veere for England.

The numbers who surrendered with General Monnet were—

| | | |
|---|---|---|
| Officers .. .. .. .. .. | 117 | Total 4,379. |
| Non-commissioned officers, rank and file | 3,773 | |
| Sick .. .. .. .. .. | 489 | |

1,000 sick and wounded had been sent to Cadzand before the surrender.

During the siege the English fleet had landed a party of seamen who manned one of the shore batteries.

It was fortunate for the besiegers that the capitulation was not delayed, as the inundations were becoming serious, and in order to check their further progress the possession of the sluice gates in Flushing had become a matter of the greatest importance.

From the time of landing to the capitulation of Flushing, the English casualties were as follows:—

| | Killed. | Wounded. | Missing. |
|---|---|---|---|
| Officers .. .. .. .. | 5 | 45 | 1 |
| Non-commissioned officers.. .. | 4 | 31 | — |
| Trumpeters and drummers .. | 3 | 3 | — |
| Rank and file .. .. .. | 105 | 507 | 43 |
| Total .. .. | 117 | 586 | 44 |

At the same time that the fall of Flushing made the English masters of Walcheren, Lord Rosslyn and Rear-Admiral Keats concluded a capitulation with the deputies from Brouwershaven and Zierikze, by which those towns and the islands of Schouwen and Duiveland surrendered.

The intelligence of the capture of these two islands was very welcome to Lord Chatham, as the distress was becoming great in Walcheren, and there was little money left in the military chest. He hoped now to be able to draw large supplies of cattle, spirits, and biscuit from Schouwen and Duiveland, which hitherto had escaped the ravages of war.

To return to the French preparations for the defence of Antwerp. On the first intimation of danger King Louis had

1809. Walcheren Expedition.

hurried from Aix-la-Chapelle to Amsterdam, and thence to Antwerp, where between the 10th and 12th August 8,000 to 10,000 men, sufficient for the garrison, had been collected. The King bringing with him his Dutch Guards, had completely denuded his own kingdom, and left Holland open to the English; but he resolutely undertook the defence of Antwerp, and wished immediately to open the dykes, inundate the surrounding country, and sink vessels in the Scheldt. Luckily for Antwerp, Decaux, the Commanding Engineer, persuaded the King to adopt the more moderate measures of confining the inundations to the neighbourhood of Forts Lillo and Liefkenshoek, and to content himself with repairing the existing works round Antwerp.

As soon as Napoleon heard of the English descent on the Scheldt he saw the importance of the blow, and issued the most energetic orders for the protection of Antwerp. He ordered Bernadotte to assume supreme command, and though not cordially approving either this appointment or Fouché's extreme measure of calling out the National Guard everywhere, considered that under the pressing circumstances of the case his minister in Paris had erred on the right side.

Bernadotte reached Antwerp on the 15th August, and immediately assumed command. By this time some 20,000 men of various corps and nations were assembled: of these, about 12,000 or 15,000, with 24 guns, could take the field.

Bernadotte made a hurried inspection of this army, and of the fortifications and defences of the Scheldt. The engineers were directed to hasten the repair of the old fort Frederick Henry, and strengthen Lillo and Liefkenshoek. The garrison was distributed as follows:—

One division at Beevendrecht occupying Santvliet, and guarding the dykes from that place to Fort Lillo.

The light cavalry was quartered at Stabroek, in order to support the division at Beevendrecht, or, in case of necessity, cover their retreat on Antwerp.

A second division was placed in *échelon* between the last division and Antwerp, parallel to the Scheldt.

The Dutch troops left by King Louis (who returned to his kingdom on Bernadotte's arrival) were placed at Woensdrecht and Ossendrecht, under the orders of Lieutenant-General Dumonceau.

The National Guard collected round Bruges, were placed at Oostburg and Ysendick as a reserve to General Rousseau's troops in Cadzand, and another body of National Guards were assembled at Hulst.

The recruits, who poured in from all directions, were incessantly drilled when not employed on the works.

By the 24th August Antwerp was secure from a *coup-de-main*. The batteries and forts along the Scheldt were in a condition to offer serious resistance to an enemy moving up the river; and to guard against the dangers of fire-ships the squadron of Admiral Missiessy was moved up above Antwerp.

In obedience to orders received from Napoleon, and dated Schönbrunn, 3rd August, 1809, the French forces in the Low

1809. Walcheren Expedition.

Countries were subsequently organised in three armies: that of Antwerp, under Bernadotte; of the Tête de Flandre, under Bessières; and a Reserve. At the same time Kellermann was ordered to move up to Maestricht.

It was still necessary to fill up the ranks of the army of the Danube, so that whilst all cadres moving to join that army were ordered to turn north, detachments passing by Strasbourg were still to continue their march towards Vienna. Napoleon looked upon the National Guard as the material for the best conscripts, as the force was entirely composed of men between wenty and thirty years of age, and he thought that, with their assistance, the regular troops who were ordered to the Scheldt would be sufficient to stop the English from advancing further. At the time, though victorious on the Danube, he could not have afforded to allow the English to threaten his flank, and was extremely anxious that prompt measures should be taken to turn them out of the Low Countries.

Lord Chatham had already lost seventeen valuable days in Walcheren, and, now that Flushing had fallen, he made no attempt to make up for lost time.

Writing from Middleburg on the 19th August, he stated that he intended to proceed to Bath the following day; but on that day he again wrote to say that the difficulty in carrying out his instructions as to the raising of money, still delayed his departure.

The 1st and 3rd Divisions, cavalry and heavy artillery, were ordered to be conveyed by water to Bath, there to join the 2nd Division reserve, and light troops. Lord Paget's Division is not mentioned, but it most likely remained with Lieutenant-General Fraser's (the 4th), under Sir Eyre Coote's command, to guard Walcheren.

Lord Chatham at length made a move to Goes, and subsequently joined Sir John Hope's reserve at Bath on the 25th August, where the whole army, except the garrison of Walcheren, was now assembled.

There were now some 23,000 infantry and 2,000 cavalry available for further operations; but in face of the numbers the French had collected between Bergen-op-Zoom and Antwerp, Lord Chatham did not dare risk an attack on the latter fortress.

On the 26th the English flotilla were off Bath, and the French were in hourly expectation of an attack. The number of small craft in possession of the English would have rendered the passage of the Bergen-op-Zoom Canal easy, even if the canal had not been fordable at low-water.*

Instead of making a bold attack, Lord Chatham assembled a Council of War consisting of the Lieutenant-Generals present with the army, and submitted for their opinion a memorandum on the

* On the 15th October, 1809, the Dutch attempted to drag a large piece of cannon over the ford near Ossendrecht, but it was lost in the mud; they subsequently, however, succeeded in getting across several guns drawn by eight horses. Cavalry also crossed, the water being only up to their horses' bellies.

Thiers also states that it was fordable at low-water, men being covered to their shoulders.

1809. Walcheren Expedition.

comparative strength of the French and English forces. The decision, like that of every other similar Council, was against action.

In a letter stating this opinion, Lord Chatham gave as his reasons for the abandonment of the object of the expedition that the enemy were assembled in too great force for his small army successfully to attack Liefkenshoek and Lillo, before moving on Antwerp; that it would be necessary to invest Breda and Bergen-op-Zoom, and to garrison Walcheren and South Beveland; and that the climate was telling on the health of his men, of whom there were already 3,000 sick.

Orders were accordingly given to strengthen the garrison of Walcheren, and embark the remainder of the troops in readiness to sail for England.

In his letter Lord Chatham quite neglected to mention the support he might have received from the navy had he decided to continue operations against Antwerp. The omission was pointed out in letters from home, and though Lord Chatham was allowed to withdraw his army, he was instructed before leaving the Scheldt to co-operate with the navy in any manner which the Admiral might consider most efficacious to close the navigation of the Scheldt, and destroy the works at Terneuse.

It was also decided to hold Walcheren, and a report was to be immediately furnished on its capabilities for defence, &c.

The sickness since known as the "Walcheren fever" had been making terrible ravages amongst the men left in that island. Sir Eyre Coote, writing from Middleburg, asked for additional medical assistance, and suggested that three or four large hospital ships should be fitted up and sent from England, as he stated that the sea air was found to do the sick good.

The state of Flushing after the bombardment was terrible: a great number of the houses had been burnt or unroofed; the inundations had produced great sickness, and the capture of the town had disorganised the local government. The want of water for drinking purposes was also becoming a serious question.

In answer to a petition from the mayor and magistrates of Flushing relative to local government, finances, &c., the English ministry decided that the inhabitants should be governed by French or Dutch laws, according to their choice. The expenses were to be defrayed from duties levied on the trade which was sure to spring up quickly between the inhabitants and the army and navy.

To supply the necessary quantity of drinking water, arrangements were made to send in casks 500 tons per week from England. This water was to be issued under the same rules as those in force on board H.M.'s ships when at sea.

As soon as Lord Chatham's decision was known, the fleet began to drop down the Scheldt; on the 30th August there were only sixty sail before Bath, and by the 4th September every English ship had cleared out of the Sarftingen Roads. South Beveland was completely evacuated, and the troops on board ship ready to sail for England, were off Flushing on the 6th September. The sickness amongst the army had increased terribly; there were now 8,000

1809. Walcheren Expedition.

men down, and the number of General officers who had gone home on sick leave, led to great embarrassment as to command. Lord Chatham on leaving Bath took up his quarters at Middleburg, waiting the orders which would permit of his returning home.

The troops he intended to leave in Walcheren numbered 825 officers, 17,845 men, and 268 horses, exclusive of staff.

The Commanding Engineer reported that the erection of the necessary barrack accommodation would cost 30,000*l.*, and the repair of the fortifications 40,000*l.*

By orders from home, the officers in charge of the civil administration of the island, before the arrival of the English, were to be retained in their posts, with the exception of French officials whose loyalty to the English Government could not, under the circumstances, be counted on.

No shipping was to be allowed either to enter or clear, unless provided with an English pass.

The daily state of the troops returning to England showed 966 officers, 23,564 men (of whom 5,025 were sick and unfit for duty), and 2,621 horses.

On the 9th September, Lord Chatham wrote from Middleburg that he was handing over the command in Walcheren to Lieutenant-General Eyre Coote, and intended to sail for England the following day.

The alarming sickness from which the army was suffering created the greatest anxiety in England, and Lord Chatham was directed to forward a report from the chief medical officer on the nature of the disease.

This was furnished by Inspector of Hospitals John Webb, whose report forms a valuable addition to the medical history of the expedition.

The following extracts will show the cause and nature of the fever:—

"The island, being so flat, and nearly level with the sea, is little better than a swamp; the ditches are filled with putrid vegetable and animal matter; the quantity of pure water very limited."

"The inhabitants are sickly and infirm."

"The sickly season begins about the middle of August, and continues till the frost stops the exhalations from the earth; the dry hot weather causing the greatest amount of sickness."

"Nearly one-third of the population is attacked with fever every sickly season, in spite of the greatest attention to cleanliness both in buildings and person."

"The fever first showed amongst the troops in South Beveland, who had not the opposition of an enemy to keep their minds and bodies in healthy action. But on the fall of Flushing it broke out amongs the troops in Walcheren."

"At first the disease appeared as a low fever, but subsequently took a form similar to jail fever. It spread with unexampled rapidity."

"No remedy could be devised to check the ravages, though means might be taken to mitigate the severity of the attacks."

"Men who have suffered from this fever have their constitutions so shattered that their physical power will for the future be materially diminished."

1809. Walcheren Expedition.

On the departure of Lord Chatham, Lieutenant-General Sir Eyre Coote assumed the command. Owing to the fever which continued to spread in an alarming manner, the situation of the new Commander-in-Chief was one of extreme difficulty.

On the 10th September there were 220 officers and 8,095 men unfit for duty. The Inspector-General of Hospitals himself was laid up with the fever.

Before assuming command, Sir Eyre Coote had drawn up a memorandum on the defence of the island, in which he pointed out the inability of the navy to keep the sea during the winter, and the consequent necessity for immediately repairing the defences and maintaining a very strong garrison.

The dykes had been very much injured during the siege of Flushing and the material for their repair was exhausted. The contractor imported the fascines and pickets, used for the repair of the dykes from Holland and Brabant, but owing to the state of war, he was unable to provide the requisite quantity. To remedy these pressing defects, Sir Eyre Coote advanced the sum of 15,000 florins from the revenues of the island seized on its capture, and granted twenty-nine blank passes to be used by the contractor's men in their journeys to and from the mainland.

The difficulties of carrying on the Government of the island were greatly increased by the inaction of the civil authorities, who dreaded the revenge of the French on the withdrawal of the English.

The garrison also was insufficient; 20,000 men had been the original estimate for the defence of the island. Only 16,000 had been detailed for this duty, and of these 8,200 were now sick. During the last fortnight the casualties had numbered 498.

Though the medical officers displayed the greatest zeal, the condition of the force was deplorable. Medical assistance had been demanded from England, but had not arrived. The hospitals established at Middleburg, Flushing, Veere, Arnemuiden, Zoutelande, and Rammekens were crowded. At Middleburg there were not sufficient beds to allow of each patient having one to himself.

From a report furnished by Deputy Inspector of Hospitals Mr. Barrow, the fever was degenerating into typhus, especially amongst the men who had lately returned from Corunna.

The daily state of the 19th September showed 224 officers, and 9,627 men sick.

On taking over the command at Walcheren from Lord Chatham, Sir Eyre Coote had expressed his anxiety to return home on private affairs, and he now put his application in an official form, as he stated that he could not much longer remain abroad.

The reiterated appeals of the General had their effect on the English Government, who made great efforts to secure sufficient accommodation for the transport of the sick and wounded. In addition to the 141 ships of 29,421 tonnage available for the purpose at Walcheren, 40 ships of 8,193 tons were taken up in various English ports, with the object of being sent to the Scheldt. With regard to medical assistance, it was a far more difficult matter, owing to the drain already caused on England by the necessities of the army in Spain.

1809. Walcheren Expedition.

The absence of a regular Government, and the destruction caused by the bombardment of Flushing, began to tell very severely on the inhabitants. The presence of a large army had naturally raised the price of all articles of food, and the disturbance of the ordinary course of civil life had deprived many of their usual means of support.

The old pensioners of the Dutch East India Company were amongst the chief sufferers, and Sir Eyre Coote applied to the home authorities for permission to pay these men their usual pensions.

On the 23rd September, information was sent to headquarters at Middleburg that all communication with South Beveland was cut off, that 4,000 Dutch troops occupied that island, and that 16,000 French were expected in Goes. It was also reported that Fort Bath had been strengthened, and that Bernadotte had recently visited it and inspected the troops in the neighbourhood.

With this intelligence the increase of sickness amongst his troops was anything but cheering to the English General, who, writing home the same day, stated that the 23rd Regiment had been so weakened by the fever that he had ordered it home at once; that the 6th and 81st Regiments were so sickly that they were struck off duty, whilst the 77th and 84th were nearly as bad. Two thousand sick were ordered home to make room in the hospitals; but the fever continued to spread, and unfortunately at the same time medical comforts began to fail.

If it were intended to hold Walcheren, Sir Eyre Coote suggested that the reinforcements should not be sent till the end of October or beginning of November, at which time he hoped that the worst of the sickly season would be over.

The doctors in Walcheren could render but little assistance to those of the English army, owing to the difficulty the men found in understanding the instructions of the Dutch doctors, and to the inferior professional attainments of the medical men themselves.

The boisterous weather, though it detained the sick and prevented their embarkation, did not check the fever, and on the 29th September there were in the 81st Regiment only forty men fit for duty.* On the 30th, Dr. Blane, Acting Physician-General, and Dr. MacGregor, Inspector-General of Hospitals, arrived from England. They reported to the home authorities that, for the conveyance of the sick and convalescents to England, the guns should be taken out of a certain number of war vessels, and that they should be fitted up with hammocks, &c.; the invalids embarked to be under the medical care of the naval surgeons belonging to the vessels used as transports. Accordingly four line-of-battle ships were fitted up as recommended, and ordered to Flushing to bring home the invalids. It is to be noted that the French doctors did not concur in this

* The following extract illustrates the rapid growth of the fever in the case of the 81st regiment:—

| | | | | |
|---|---|---|---|---|
| July, '09, strength embarked | 656 | men fit for duty. | | |
| 7th Sept., '09, Walcheren, | „ | 468 | „ | „ |
| 29th Sept., '09, Walcheren, | „ | 40 | „ | „ |

1809. Walcheren Expedition.

method of dealing with the fever-stricken patients; for while the English surgeons recommended the removal of the sick on board ship as their only chance of recovery, General Monnet had reported to his Government that the men attacked with fever recovered more quickly if kept on the island than if removed from it, either on board ship or to the mainland.

At the outset of the expedition fifty men of the 7th Royal Veteran Battalion had been assigned to it as hospital orderlies, and assistants to the sick. On the great increase in the number of the patients,* Sir Eyre Coote had applied for 300 more men to be sent for service in the hospitals; this application was refused, and the General was instructed to get his assistants from amongst the natives at the rate of ten men per regiment. Accordingly, on the 6th of October, an order was issued to that effect, which also authorised the payment of one florin per diem to these native assistants, who were termed Regimental Pioneers.

This day (6th October), the 23rd Regiment started from Middleburg to embark at Flushing, and being unable to march were conveyed in waggons. A large number of convalescents were also removed to the hospital ships "Asia" and "Britannia" for change of air.

A barrack department was established on the island, some 400 Dutch being employed under it repairing buildings, &c. At this period the troops were unprovided with warm clothing for the winter, and had no heavy baggage with them, and, to add to their other discomforts, no cook-houses had been built for their use.

A petition from the inhabitants was presented to Sir Eyre Coote to disallow a claim of the Artillery to take away the bells of the church of Flushing, which they asserted was their right, in the case of a town captured by siege. The Artillery waived their claim on the inhabitants promising to pay 2,000*l.*, a sum which was subsequently reduced to 500*l.* on account of the losses the townspeople had incurred during the siege, and the previous occupation of the French.

On the 26th October, from a return of medical officers doing duty with the eighteen line regiments, out of a proper complement of fifty-four only twenty-three were fit for duty. The average number of sick per regiment was 400 men.

Sir Eyre Coote now sent an application for reinforcements, as this was the most favourable time of year for troops to arrive on the island with the hope of becoming acclimatised; and added that from deaths, sickness, and the number sent home, the garrison of Walcheren could only muster 4,000 men fit for duty. About this time Captain Owen, H.M.S. "Pallas," reported an impending attack, as the enemy had recently been reinforced, and were collecting large numbers of boats at Bergen-op-Zoom and Goes with the apparent intention of crossing the various channels.

| | Officers. | Serjeants. | Drummers. | Rank and File. |
|---|---|---|---|---|
| * Daily State of Sick, 6 Oct., 1809. | 169 | 334 | 156 | 8,855 |

1809. Walcheren Expedition.

The weather for the last three weeks had been very fine, but this had no effect in decreasing the numbers of sick, and although during that time 4,000 had been invalided to England, the number of deaths was 128.

Sir Eyre Coote now resigned his command to Lieutenant-General Don, who had arrived on the 24th October.

On taking over command, General Don reported to the Government that there were over 4,000 sick, and urgently demanded transports for their removal. As some time must elapse before the ships could arrive, he estimated the number by the date of their probable arrival at over 5,000. Of the men fit for duty (4,534), after deducting orderlies, servants, &c., he stated that not one-third were in a fit state to march five miles rapidly, or to perform the duties of night patrols. At this time the Government had not come to a decision as to whether they would hold the island of Walcheren or evacuate it, after destroying the docks and defences. With a view to the former course being adopted, some 5,000 men were collected on the east coast of England ready to reinforce the garrison, and with a view to giving up the island instructions were sent to General Don to make all necessary preparations for destroying its defences as well as the two basins in the docks at Flushing.

In answer to inquiries from the Government as to how long the latter operations would take, General Don estimated that to destroy the docks of Flushing would require six days, but that if all the works on the island as well as the docks were destroyed twenty days must elapse before the garrison could be embarked.

On receiving the announcement that a force of 5,000 men was ready to embark for Walcheren to reinforce him, he wrote to say that if he was to hold the island such an addition of force was not sufficient, and that if attacked whilst embarking he would inundate the country round Middleburg, and thus confine the advance of the enemy to the dykes.

In reply to questions from the Government on the force necessary to hold the island of Walcheren, General Don sent in a statement to the following effect:—

The island is about thirty-four miles in circumference. Owing to the severity of the winter, the ice prevents any co-operation in its defence by a naval force during the winter months. It is open to attack from the north side round by the east to its south-west side, from the following points:—the East Scheldt, North Beveland, Wolfaartsdij, South Beveland, the West Scheldt, Ostend, and Hellevoetsluis. In order to hold the island, it would be necessary to fortify Flushing and Veere, and to construct casemates in those places for 3,500 and 1,500 men respectively.

Bomb-proof hospitals and stores would also be required. Military stations at various points would be a necessity, in addition to twenty-eight Martello towers, mounting from one to five heavy guns each, for the protection of exposed portions of the coast.

He estimated the strength of the garrison for permanent defence at:—

1809. Walcheren Expedition.

12 brigades of Field Artillery.*

950 { 200 Artillerymen for garrison of Flushing,
100 " " Ter Veere,
250 " " Martello towers,
400 " " Coast batteries, }

7 troops of cavalry,

21,000 infantry,

making in all a force of 23,150 men.

He calculated the strength of the enemy in Antwerp at 22,100 men and at 7,000 in West Beveland, the former force composed chiefly of French troops, the latter of Dutch: and from the fact that two battalions of the Imperial Guard had arrived at Antwerp he assumed the speedy arrival of Napoleon.

In conclusion, General Don stated that at that time the island was in a defenceless state; that his forces were unable to act in the field; and that as there was no bomb-proof cover in Flushing he could only hold that place until the enemy commenced their mortar fire.

Orders were now received to destroy the works and basins of Flushing preparatory to embarking.

In order to take in flank an advance from South Beveland, General Don applied for 1,800 men, and four companies of artillery of the force assembled on the east coast of England, who on arrival were to be placed in flat-bottomed boats in the Veersche Gat.

The interval between the 7th and 20th November was employed in preparations for destroying the works and docks of Flushing and other places on the island: and in completing the embarkation of the sick on board the transports which had now arrived from England.

On the night of the 1st November the boats of the advance guard flotilla in the West Scheldt cut out some of the enemy's boats at Hoedekenskerke, which carried ten iron 36-pounders, and eight brass 18-pounders, together with carriages and platforms.

The enemy were reinforced by some 1,200 men in North Beve-

* At this time a field brigade of artillery consisted of six pieces of ordnance (guns and howitzers), and was composed as follows, viz.:—

| | Men. | | | | | Animals. | Carriages. |
|---|---|---|---|---|---|---|---|
| | Officers. | N.C. officers. | Gunners. | Drivers. | Artificers. | | |
| Company of Artillery .. | 5 | 17 | 123 | — | — | 160 horses | 19 |
| Driver Corps .. .. | 1 | 9 | — | 96 | 10 | 10 mules | — |

land, and had brought some heavy artillery to north-west of South Beveland to drive the English flotilla out of the various channels of the Scheldt.

1809. Walcheren Expedition.

The flood-gates of Flushing basin were destroyed, but it was found that the mining operations would take longer than estimated and the embarkation delayed in proportion.

The greater portion of the troops were embarked, rear guards being left at Veere, Rammekens, and Middleburg, the fresh water sluices being opened to restrict enemy's advance to the dykes.

The weather, which had been boisterous for the last few days, now became worse, and several of the smaller transports were driven ashore.

The guns of the flotilla in Wolfaartsdijk silenced the enemy's batteries.

The fleet was now ready to sail, the ships being divided into four divisions, according to the port at which they had to land their troops on reaching England: 1st division in front of Flushing, 2nd division off the coast to westward of Flushing; 3rd between Flushing and Rammekens; 4th off Veere.

The 3rd division to co-operate with the navy in the Sloe Passage; the 4th to act between Veere and Wolfaartsdijk in case of an attack.

The rear guards were then withdrawn from Veere; Rammekens, and Middleburg.

With a view to the necessity of again disembarking to repel an attack, two days' provisions were kept ready cooked, each man was provided with 60 rounds of ammunition, 6 flints, 1 blanket, 1 pair of shoes, and 1 pair of socks. The packs in that case, to be left on board ship.

As the enemy made no attempt to molest the departing expedition, there was no necessity for disembarking the troops; and in the course of a few more days the several divisions sailed for England, where all arrived before the close of the year.

This expedition, which effected so little, cost the country about 835,000*l.* sterling, and the lives of some 4,000 men, of whom only 106 fell in battle.

Out of a total strength of over 35,000 officers and men who landed with the returning expedition, more than 11,500 were in hospital, of whom numbers subsequently died, and of those who recovered the greater portion carried to their early graves a ruined constitution, the legacy left by the fever fens of Walcheren.

---

## APPENDIX A.

Appendix A. Naval Force.

*Amount of Naval Force.*

35 Sail of the Line.
2 Ships of 50 Guns.
3 Ships of 44.
18 Frigates.

1809. Walcheren Expedition. Appendix A. Naval Force.

33 Sloops.
5 Bomb Vessels.
23 Gun Brigs—five carrying mortars.
17 Hired Cutters.
14 Revenue Vessels.
5 Tenders.
82 Gunboats.

Together with the craft employed in His Majesty's Dockyards.
*Admiralty, 11th July, 1809.*

---

Composition and strength of Force embarked.

RETURN of the regiments now under orders for Foreign Service, showing the probable number of Rank and File, which will embark with each corps, leaving behind such men as are at present unfit for duty.

ADJUTANT-GENERAL'S OFFICE, 15*th July*, 1809.

Adjutant-General's Return.

| | | | |
|---|---|---|---|
| Cavalry | 3rd Dragoons.. .. .. | 6 troops .. | 510 |
| | 9th Light Dragoons .. .. | ,, .. | 510 |
| | 12th ,, ,, .. .. | ,, .. | 510 |
| | 2nd German Light Dragoons | ,, .. | 580 |
| | 3rd ,, ,, ,, | 2 ,, .. | 152 |
| | Waggon Train .. .. | 5 ,, .. | 395 |
| | | | 2,657 |
| Artillery.. | Horse Artillery | 1 troop .. .. .. | 150 |
| | Foot ,, | 16 companies .. .. | 1,839 |
| | Gunner Drivers | .. .. .. .. | 1,043 |
| | | | 3,032 |
| Foot Guards | 1st Foot Guards | 1st battalion .. .. | 1,329 |
| | ,, | 3rd ,, .. .. | 1,101 |
| | Flank Companies | .. .. .. .. | 437 |
| | | | 2,867 |

*Infantry.*

| | | | |
|---|---|---|---|
| Royals .. | .. .. | 3rd battalion .. .. | 957 |
| 2nd Foot | .. .. | .. .. .. .. | 833 |
| 4th ,, | .. .. | 1st ,, .. .. | 1,000 |
| ,, ,, | .. .. | 2nd ,, .. .. | 930 |
| 5th ,, | .. .. | 1st ,, .. .. | 939 |
| 6th ,, | .. .. | 1st ,, .. .. | 971 |
| 8th ,, | .. .. | Two companies.. .. | 200 |
| 9th ,, | .. .. | 1st battalion .. .. | 932 |
| 11th ,, | .. .. | 2nd ,, .. .. | 839 |
| 14th ,, | .. .. | 2nd ,, .. .. | 781 |
| 20th ,, | .. .. | .. .. .. .. | 873 |
| 23rd ,, | (4 companies).. | 2nd ,, .. .. | 400 |
| 26th ,, | .. .. | 1st ,, .. .. | 687 |
| 28th ,, | .. .. | 1st ,, .. .. | 650 |
| 32nd ,, | .. .. | 1st ,, .. .. | 579 |
| 35th ,, | .. .. | 2nd ,, .. .. | 737 |
| | | Carried forward | 12,308 |

1809. Walcheren Expedition. App. A, Composition and strength of Force embarked.

| | | | |
|---|---|---|---|
| | | Brought forward | 12,308 |
| 36th Foot | .. .. | 1st battalion .. .. | 657 |
| 38th " | .. .. | 1st " .. .. | 793 |
| 42nd " | .. .. | 1st " .. .. | 799 |
| 43rd " | .. .. | 2nd " .. .. | 604 |
| 50th " | .. .. | 1st " .. .. | 853 |
| 51st " | .. .. | .. .. .. .. | 652 |
| 52nd " | .. .. | 2nd " .. .. | 418 |
| 59th " | .. .. | 2nd " .. .. | 740 |
| 63rd " | .. .. | 2nd " .. .. | 400 |
| 68th " | .. .. | .. .. .. .. | 777 |
| 71st " | .. .. | 1st " .. .. | 963 |
| 76th " | .. .. | .. .. .. .. | 742 |
| 77th " | .. .. | .. .. .. .. | 559 |
| 79th " | .. .. | .. .. .. .. | 1,003 |
| 81st " | .. .. | 2nd " .. .. | 661 |
| 82nd " | .. .. | 1st " .. .. | 1,000 |
| 84th " | .. .. | 2nd " .. .. | 855 |
| 85th " | .. .. | .. .. .. .. | 581 |
| 91st " | .. .. | 1st " .. .. | 660 |
| 92nd " | .. .. | 1st " .. .. | 987 |
| 95th " | .. .. | 2nd " .. .. | 1,000 |
| Staff Corps | .. .. | 2 companies .. .. | 100 |
| 1st German Light Battalion | | .. .. .. .. | 704 |
| 2nd " " | | .. .. .. .. | 613 |
| Embodied Detachments | | .. .. .. .. | 800 |
| | | | 31,229 |

*Abstract.*

| | |
|---|---|
| Cavalry .. .. .. .. .. | 2,657 |
| Artillery .. .. .. .. .. | 3,032 |
| Foot Guards .. .. .. .. | 2,867 |
| Infantry .. .. .. .. .. | 31,229 |
| | 39,785 |
| Increase of Cavalry .. .. .. | 510 |
| | 40,295 |
| Deduct two troops withdrawn .. .. | 152 |
| Total .. .. | 40,143 |

N.B.—A regiment of Dragoons will be named for service instead of the two troops of the 3rd German Light Dragoons.

It is expected that the regiments will embark something stronger than is here stated.

(Signed) HARRY CALVERT,
*Adjutant-General.*

NOTE.—2nd Dragoon Guards afterwards sent instead of 3rd German Light Dragoons.

1809.
Walcheren Expedition.
Return showing casualties on 7th Sept.

# APPENDIX B.

RETURN of the several Corps to remain in the Island of Walcheren. *Middlebourg, 7th September*, 1809.

| Brigades. | Regiments. | Officers present. | | | | Quarter-masters. | Serjeants. | | Trumpeters or Drummers. | | Rank and File. | | | Horses. | | |
|---|---|---|---|---|---|---|---|---|---|---|---|---|---|---|---|---|
| | | Field Officers. | Captains. | Subalterns | Staff. | | Present fit for duty. | Sick. | Present fit for duty. | Sick. | Present fit for duty. | Sick. | Total. | Present fit for duty. | Sick. | Total. |
| | 9th Light Dragoons ... | — | 1 | 2 | — | — | 4 | — | 1 | — | 80 | — | 80 | 80 | — | 80 |
| | Royal Artillery ... | 3 | 16 | 28 | 10 | — | 22 | — | 14 | — | 1,036 | — | 1,036 | — | — | — |
| | Royal Engineers* ... | — | — | — | — | — | — | — | — | — | — | — | — | — | — | — |
| Brigadier-General Kettenberg | 68th Foot ... | 1 | 7 | 22 | 5 | — | 18 | 20 | 19 | 3 | 439 | 316 | 755 | — | — | — |
| | 71st ,, 1st battalion | 2 | 6 | 25 | 4 | — | 28 | 20 | 16 | 5 | 603 | 339 | 942 | — | — | — |
| | 85th ,, ... ... | 1 | 6 | 29 | 4 | — | 28 | 11 | 14 | 4 | 437 | 132 | 569 | — | — | — |
| Brigadier-General Allen | 1st Lt. Batt. K.G.L. ... | 3 | 3 | 14 | 2 | — | 31 | 6 | 13 | 3 | 477 | 227 | 704 | — | — | — |
| | 2nd ,, ,, ... | 3 | 3 | 14 | 3 | — | 24 | 9 | 11 | 5 | 405 | 205 | 610 | — | — | — |
| | 1st Foot, 3rd battalion | 3 | 8 | 20 | 6 | — | 41 | 10 | 18 | 3 | 784 | 150 | 934 | — | — | — |
| Colonel Hay | 5th ,, 1st ,, | 2 | 7 | 24 | 6 | — | 22 | 15 | 10 | 6 | 483 | 431 | 914 | — | — | — |
| | 35th ,, 2nd ,, | 1 | 5 | 17 | 6 | — | 17 | 22 | 11 | 8 | 420 | 310 | 730 | — | — | — |
| Brigadier-General Ackland | 2nd ,, | 2 | 7 | 21 | 5 | — | 35 | 8 | 15 | — | 608 | 217 | 825 | — | — | — |
| | 76th ,, | 3 | 10 | 27 | 6 | — | 37 | 4 | 13 | 3 | 544 | 189 | 733 | — | — | — |
| | 32nd ,, 1st ,, | 3 | 7 | 21 | 4 | — | 34 | 6 | 11 | 9 | 366 | 192 | 558 | — | — | — |
| Major-General Dyott | 6th ,, 1st ,, | 2 | 7 | 26 | 6 | — | 24 | 29 | 13 | 9 | 450 | 496 | 946 | — | — | — |
| | 50th ,, 1st ,, | 2 | 6 | 27 | 6 | — | 26 | 18 | 14 | 6 | 523 | 332 | 855 | — | — | — |
| | 91st ,, 1st ,, | 3 | 6 | 25 | 6 | — | 30 | 12 | 14 | 1 | 431 | 205 | 636 | — | — | — |
| Brigadier-General Brown | 23rd ,, 2nd ,, | 1 | 5 | 11 | 5 | — | 8 | 11 | 7 | 4 | 154 | 244 | 398 | — | — | — |
| | 81st ,, 2nd ,, | 1 | 8 | 25 | 5 | — | 27 | 13 | 10 | 8 | 468 | 187 | 655 | — | — | — |
| | 84th ,, 2nd ,, | 3 | 7 | 24 | 5 | — | 23 | 16 | 11 | 3 | 548 | 236 | 784 | — | — | — |
| | 26th ,, 1st ,, | 2 | 8 | 26 | 5 | — | 40 | 8 | 14 | 2 | 523 | 129 | 652 | — | — | — |
| Brigadier-General Montresor | 36th ,, 1st ,, | 3 | 6 | 24 | 6 | — | 34 | 12 | 12 | 5 | 420 | 269 | 689 | — | — | — |
| | 38th ,, 1st ,, | 3 | 8 | 29 | 5 | — | 31 | 16 | 11 | 5 | 453 | 352 | 805 | — | — | — |
| | 77th ,, | 2 | 5 | 19 | 5 | — | 15 | 19 | 13 | 4 | 262 | 284 | 546 | — | — | — |
| | Royal Staff Corps ... | — | 1 | 2 | — | — | 2 | — | 1 | — | 50 | — | 50 | — | — | — |
| | Royal Waggon Train... | 1 | 2 | 6 | 2 | — | 9 | — | 2 | — | 160 | — | 160 | 188 | — | 188 |
| | Total ... ... | 50 | 155 | 508 | 117 | — | 610 | 285 | 288 | 96 | 11,124 | 5,442 | 16,566 | 268 | — | 268 |

* The detachment of this corps to remain cannot be stated at present.

The above is the strength of the different corps, agreeable to the last Weekly State, dated the 3rd instant. The casualties that have since taken place are, therefore, not accounted for; but the columns of sick are filled up from the reports of this morning.

(Signed) ROBERT LONG, Colonel, *Adjutant-General.*

## APPENDIX C.

1809.
Walcheren
Expedition.
App. C,
Return of sick
and wounded
sent to England at
various times.

RETURN of the Sick and Wounded, sent to England from Walcheren, at different times.

|  | |  |  |
|---|---|---|---|
| 21st August.. | From | Veere .. .. | 200 |
| 9th September | .. | do. .. .. | 75 |
| 16th „ | .. | Rammekens .. | 75 |
| 18th „ | .. | Veere .. .. | 334 |
| 5th October | .. | Flushing.. .. | 971 |
| 9th „ | .. | do. .. .. | 802 |
| 10th „ | .. | do. .. .. | 738 |
| 21st „ | .. | Veere .. .. | 397 |
| 22nd „ | .. | Flushing.. .. | 1,001 |
| 28th „ | .. | do. .. .. | 170 |
| 30th „ | .. | do. .. .. | 705 |
| 31st „ | .. | do. .. .. | 681 |
| „ „ | .. | do. .. .. | 250 |
| „ „ | .. | Veere .. .. | 111 |
| 11th November | .. | do. .. .. | 761 |
| „ „ | .. | Flushing.. .. | 369 |
| 14th „ | .. | do. .. .. | 520 |
| 15th „ | .. | do. .. .. | 336 |
| 17th „ | .. | do. .. .. | 600 |
| 23rd „ | .. | do. .. .. | 1,173 |
| 26th „ | .. | do. .. .. | 680 |
| 28th „ | .. | do. .. .. | 144 |
| 1st December | .. | do. .. .. | 149 |
| 14th „ | .. | do. .. .. | 184 |
| 16th „ | .. | do. .. .. | 48 |
| 31st August<br>4th September | .. | South Beveland.. | 1,389 |
| | | Total .. .. | 12,863 |

1809. Walcheren Expedition. App. D.

## APPENDIX D.

RETURN showing the Effective Strength of the Army, which embarked for service in the Scheldt, in the month of July, 1809; the Casualties which occurred; the Number of Officers and Men who returned to England, and the number reported Sick, according to the latest returns (with the exception of the 59th Regiment, from which Corps a proper return has not yet been received).

ADJUTANT-GENERAL'S OFFICE, 1*st February*, 1810.

| | Officers. | Rank and File. | Officers. | Serjeants, Trumpeters, Drummers, and Rank and File. |
|---|---|---|---|---|
| Embarked for service .. .. .. .. .. | | | 1,738 | 37,481 |
| Killed .. .. .. .. .. .. | 7 | 99 | 67 | 4,108 |
| Died { on service .. .. .. .. | 40 | 2,041 | | |
| Died { since sent home .. .. .. | 20 | 1,859 | | |
| Deserted .. .. .. .. .. .. | — | 84 | | |
| Discharged .. .. .. .. .. | — | 25 | | |
| Total officers and men who returned, and who are now borne on the strength of their respective corps .. | | | 1,671 | 33,373 |
| Of which number are reported sick .. .. .. | | | 217 | 11,296 |

(Signed) HARRY CALVERT,
*Adjutant-General.*

1813-14. Expedition to Bergen-op-Zoom.

## EXPEDITION TO HOLLAND, 1813-14 (BERGEN-OP-ZOOM).

IN the year 1813 the people of Holland determined to throw off the French yoke, and regain their national independence under the rule of the Prince of Orange. The British Government having resolved to aid the Dutch in this attempt, a force for the purpose was despatched to the coast of Holland, under the command of Sir Thomas Graham, afterwards Lord Lynedoch.

Troops were embarked at various times during the months of November and December, 1813, with this view, at Ramsgate, Dover, and Harwich, arriving in Holland by Brigades, single battalions, or detachments, at the following places:—Scheveling (or Scheveningen), Hellevoetsluis, Stevenisse (Island of Tholen), and Willemstadt.

By the 27th of December, 1813, the different portions of the troops were concentrated in cantonments about Klundert, Willemstadt, and Zeevenbergen; the force being organised in Brigades, as follows:— 1813. Expedition to Bergen-op-Zoom.

| Brigade | Regiment | | Nominal strength. |
|---|---|---|---|
| Guards' Brigade, Major-General Cooke. | 1st Regiment | | 1,600 |
| | 2nd " | (Coldstreams) | |
| | 3rd " | " | |
| Light Brigade, Major-General Mackenzie. | 35th " | 2nd battalion | 1,900 |
| | 52nd " | 2nd " | |
| | 73rd " | 2nd " | |
| | 95th " | 3rd " | |
| 1st Brigade, Major-General Skerrett. | 37th " | 2nd " | 2,500 |
| | 44th " | 2nd " | |
| | 55th " | | |
| | 69th " | 2nd " | |
| | 1st Veteran Battalion | | |
| 2nd Brigade, Major-General Gibbs. | 25th Regiment, 2nd battalion | | 2,060 |
| | 33rd " | | |
| | 54th " | | |
| | 56th " | 3rd battalion | |
| | 5 companies of Artillery | | 615 |
| | Total | | 8,675 |

*Note.*—The battalions of the 25th, 33rd, 54th, and 73rd regiments had just returned from Swedish Pomerania, under Major-General Gibbs, and on reaching Yarmouth Roads were sent to Holland without landing.

The officers of the force were for the most part young and inexperienced men, whilst the greater portion of the rank and file were mere boys unfit for the hardships of a campaign.

During the first week of the new year the 2nd King's German Hussars, and on the 10th January the 2nd battalions of the 21st and 78th Regiments, arrived in Holland, the 21st being placed in the 1st Brigade, replacing the Veteran Battalion, which was left as a garrison at various places, and the 78th in the 2nd Brigade. 1814.

General von Bülow, who commanded the 3rd Prussian Corp d'Armée, the most advanced portion of the armies about to invade France, had his headquarters at Bommel.

The French held various fortified towns in Holland and Flanders, amongst others Bergen-op-Zoom and Antwerp; they also had troops from Wustwezel, north-east of the latter place, on the Breda-Antwerp road, by Hoogstraeten, Turnhout, to Eindhoven.

General von Bülow, at the beginning of January, determined to advance southwards from Breda, breaking through the French line at Hoogstraeten and Wustwezel, thus cutting off the detachments of the enemy to the eastward of these places; and further to make a reconnaissance of Antwerp, and carry it if possible by a *coup de main.*

Sir Thomas Graham, to assist in this operation, was to advance southwards on the right of the Prussian line of attack, and, if possible, get between the French at Hoogstraeten and the fortress of Antwerp.

The British headquarters had been advanced as far south as Calmhout (or Calmpthout), a little west of the Breda-Antwerp road.

A force of infantry some 1,200 strong, with a portion of the 2nd King George's Hussars, was left at Wouw to observe Bergen-op-Zoom.

After providing garrisons for Tholen and Willemstadt, the available force for an advance numbered about 4,500 infantry, two squadrons of the 2nd King George's Hussars, and two field batteries (twelve guns).

1814. Expedition to Bergen-op-Zoom.

The army was now organised in two Divisions, each having a battery of artillery attached to it, as follows:—

| | |
|---|---|
| 1st Division, Major-General Coote. | Guards' Brigade, Colonel Lord Proby. 1st Brigade, Major-General Skerrett. |
| 2nd Division, Major-General Mackenzie. | Light Brigade, Major-General Gibbs. 2nd Brigade, Major-General Taylor. |

On the morning of the 11th January the Prussians advanced and drove back the French from about Hoogstraeten and Westwezel as far as Braeschaet, on the Breda-Antwerp road. During the night the French drew back their line closer to Antwerp, their left resting on the village of Merxem.

The British and Prussians continued to push on their advance; the former, on the evening of that day, having the 1st Division at Capelle, the 2nd at Eckeren in front of Merxem.

The Prussians had reached Braeschaet the same night, their right having advanced by the Breda-Antwerp road, the British keeping to the west of that road.

On the 13th the French were driven in on the fortress by a combined attack of the British and Prussian troops.

The English share of the engagement was the capture of the village of Merxem, which was carried by the 2nd Division after some sharp fighting.

The loss suffered by the division was: killed, 1 officer and 9 rank and file; wounded, 4 officers and 25 rank and file. 25 rank and file of the enemy were taken prisoners.

General von Bülow, finding the fortress too strong to be taken by a *coup de main*, and not deeming it practicable to invest the place on account of the severity of the weather, determined to retire again into cantonments in and about Breda.

The British also retired northwards, the 1st Division being cantonned about Eschen and Nispen, the 2nd at Calmhout.

Towards the end of the month, General von Bülow acceded to the desire of Sir T. Graham to make another advance on Antwerp, and to attempt to destroy, by a bombardment, the fleet lying in the basin there.

On the 31st January, 1814, the Prussian headquarters had been moved to Westmalle; the British troops, having a strength of about 6,000 infantry, had their headquarters at Brecht, the 1st Division being at Wustwezel and Louenhout, the 2nd at Brecht.

The British troops had advanced as far south as Braeschaet on the evening of the 1st February, and on the morning of the next day the village of Merxem was again attacked and carried, although it had been considerably strengthened by the French since the former attack on it.

The village was carried by the 2nd Division, who captured there two guns and a few prisoners.

Under cover of his own troops in the front, and supported by the Prussians on the left, Sir T. Graham proceeded to erect batteries for the bombardment of the fleet.

1814. Expedition to Bergen-op-Zoom.

By the afternoon of the 3rd February the batteries opened fire, being armed with twelve English pieces (24-pounder guns), howitzers, and mortars, and thirteen Dutch mortars of various calibres.

This day's fire disabled most of the Dutch mortars.

At noon on the 4th, fire was again commenced with seventeen pieces of artillery; and was resumed on the 5th with the same number of guns, the bombardment lasting till sunset.

During these three days' bombardment, the ships, having had their decks covered with timber and turf, were not much damaged, except about the spars, two only of them being disabled.

The French made several attempts to carry the batteries, coming from the east of the fortress, but were driven back by the Prussian troops. The losses suffered by the British at the second taking of Merxem, and during the three days' bombardment, were: killed, 9 rank and file; wounded, 17 officers, 11 serjeants, 4 drummers, and 169 rank and file.

The enemy lost 180 prisoners to the British.

Whilst this bombardment was going on, General von Bülow received orders to advance southwards into France to co-operate with the Grand Army, then entering France from the south and east.

On 6th February Sir Thomas Graham withdrew his troops and guns from their positions in front of Antwerp, and retired northwards, again into cantonments.

On the 10th his forces were disposed as follows:—

Headquarters at Groot Zundert.

1st Division about Rozendaal.

2nd Division about Groot Zoondert, having two battalions and two guns at Louenhout and Wustwezel. The cavalry (three squadrons 2nd King George's Hussars) being spread out from Wouw and Eschen on the west, to Louenhout and Brecht on the south.

After the advance of General von Bülow in the direction of the French frontier, Sir Thomas Graham was not in sufficient force to attempt anything by himself against Antwerp, and therefore withdrew his line closer to Willemstadt, where all his stores, &c., were deposited, with the object of waiting an opportunity to undertake operations against the fortress of Bergen-op-Zoom, the occupation of which by the French prevented him moving southwards to co-operate with a force of Prussians and Saxons, then occupying the south of Holland, and which had advanced as far as Brussels.

In the beginning of February the 3rd King George's Hussars, the 4th battalion Royal Scots, and the 2nd battalion 91st Regiment arrived in Holland, having marched from Stralsund, and the 2nd battalions of the 30th and 81st Regiments landed from England, making an increase of force of 2,100 bayonets.

At this period, the French having again appeared in force about Courtrai, the Duke of Saxe-Weimar, commanding a corps of Saxon troops then investing Maubeuge, informed Sir Thomas Graham that a corps of some 3,000 infantry and a body of cavalry had started from Cambrai to reinforce Antwerp; the latter, therefore, moved the right of his corps, on the 4th March, southwards to Stabroek and Putten, to prevent any reinforcement reaching Bergen-op-Zoom from Antwerp, and fixed his headquarters at Calmpthout.

**1814. Expedition to Bergen-op-Zoom.**

The arrival of a body of Russian seamen to garrison the Island of Tholen, &c., set free some of Sir Thomas Graham's force hitherto engaged on that duty.

On the night of the 8th March an assault was made on the fortress of Bergen-op-Zoom, of which place the following is a description, taken from Cust's "Wars of the Nineteenth Century."

"Before the gate of Antwerp is a large redoubt joining the fortified lines called Kijkin-de-Pot, strengthened by four flanking forts armed with cannon. On the side of the Steenbergen are the forts of Moermont, Pinsen and Rover, with a well-fortified line of connection, beyond which is an inundation reaching all the way to the Steenbergen. Before the Water-gate is a regular fort of five bastions called Zuyd Schants, under cover of which two canals lead from the Schelde and form the harbour. On the east towards Breda is another considerable inundation caused by the waters of the Zoom, which renders the whole approach on that side marshy and inaccessible. The body of the place is defended by a rampart about a league in circumference, flanked by ten bastions, covered by six hornworks; in addition to which an extensive system of mines and subterraneous galleries render every approach to the fortress hazardous in the extreme."

Bergen-op-Zoom was a strongly fortified place, but having a very weak garrison, 2,800 men only, the length of its line of defences was a source of weakness; moreover, the ditches being unrevetted, and the scarps having crumbled away from the effects of the frost, made the passage of them comparatively easy.

Sir Thomas Graham, learning from some Dutch officers who had recently been in the place, and were well acquainted with it and its garrison, that it could be entered at three points with ease, and that the *morale* of the garrison was very poor, determined to make an effort to carry the place by assault.

The attack was organised in four columns, taken from the troops of the 1st Division; the 2nd Division to furnish a supporting force, and to observe Antwerp during the carrying out of the plan.

The first column was made up of—

600 stormers.
400 supports.

Total, 1,000 men, from the Guards' brigade, under the command of Colonel Lord Proby.

This column was ordered to move from Hoogerheide to Borgvliet, and from thence to enter the place between the southern or Antwerp gate, and the north-western or Water-gate. On gaining the ramparts, the column was to move to its left to form a junction with No. 4 column, which was to enter by the Water-gate.

The second column consisted of—

| | | | |
|---|---|---|---|
| 55th Regiment, | | 250 | stormers. |
| 69th | „ | 350 | |
| 33rd | „ | 600 | supports. |

Total, 1,200 men, under the command of Lieutenant-Colonel Morrice, 69th Regiment.

# EXPEDITION TO HOLLAND 1813-14.

Intelligence Br. No. 309.

Lithod. at the Intelligence Br. War Office Aug. 1883.

This column was to move from Huibergen to the north of the eastern or Wouw gate, with orders on gaining the ramparts to communicate to its left with No. 1 column, and to act according to circumstances. 1814. Expedition to Bergen-op-Zoom.

The third column was destined only to make a false attack, ready to co-operate with the others if successful, and consisted of—

| | |
|---|---|
| 21st Regiment, | 100 |
| 37th „ | 150 |
| 91st „ | 400 |

Total, 650 men, under command of Lieutenant-Colonel Henry, 21st Regiment. This column was to advance from Halsteren towards the north or Steenbergen gate.

The fourth column consisted of—

| | | | |
|---|---|---|---|
| | 44th regiment, | 300 | stormers. |
| Flank companies | 21st „ / 37th „ | 200 | |
| | 1st Royal Scots, | 600 supports. | |

Total, 1,100 men, under command of Lieutenant-Colonel Carleton, 44th regiment. This column was to move from Halsteren by the junction of the two dykes to the north-west of the town, and fording the Zoom stream (widened and deepened at this part into a tidal canal, passable at low water), to enter the place by the Water-gate.

On gaining the ramparts it was to move to its right and form a junction with No. 1 column.

The total strength of the four columns was 3,950 rank and file.

Each column was guided by a selected officer, having a party of sappers and miners under his command.

Major-General Cooke accompanied the first column, Major-General Skerrett and Brigadier-General Gore being with No. 4 column.

In order to distinguish friends from foes in the darkness and confusion, the men were instructed on the approach of any one to call out "Orange Boven" (Up with the Orange) the answer to which challenge was to be "God save the King."

As No. 4 column had to ford the Zoom, and it was low water about 10 P.M. on the 8th March, that hour was chosen for the simultaneous attack of all the columns.

The false attack, No. 3 column, opened fire between 9 and 10 P.M., and took the guard by surprise, but was stopped at the drawbridge by the fire from the place, and retired, suffering severely. The sound of musketry, nevertheless, attracted the greater portion of the garrison to that side.

No. 4 column was the first to enter the place, forcing the Water-gate and clearing the ramparts of the few enemy found there; instead, however, of moving to the right, according to their instructions, they scattered right and left along the ramparts on either side of the gate. While in this disorder they were attacked by the greater portion of the garrison under the commandant, General Bizanet, and were very severely handled, General Gore and Colonel

1814. Expedition to Bergen-op-Zoom.

Carleton being killed, and General Skerrett receiving a mortal wound. On the fall of their leaders the troops fell into entire confusion, and were made prisoners by the French.

The first column was delayed in its attack by having to change its direction at the last moment, as it was found impracticable to pass the ditch at the indicated spot. However, about 11.30 P.M. the column got into the Orange bastion to the westward of the Antwerp gate, under a galling fire from the few defenders at that part.

On reaching the rampart General Cooke halted, and instead of moving to the left in accordance with his orders, sent only a strong patrol to gain intelligence of the fourth column, but the patrol unfortunately fell into the hands of the enemy. A few houses in front of the column were occupied, and a detachment of the 1st Guards, under Lieutenant-Colonel Clifton, was sent to the right to attempt to open the Antwerp gate, and gain intelligence of No. 3 column; this detachment also fell into the hands of the enemy, who got between it and the Orange bastion.

A detachment of the 3rd Guards, under Lieutenant-Colonel Rooke, was shortly afterwards sent on a similar errand, but could not force its way up the street leading to the Antwerp gate, and found also that an outwork commanding the bridge at that gate was occupied by the enemy, the fire from which would have prevented any entrance at that point.

The second column was arrested in its advance at the glacis by the fire from the main body of the place: its leader, Lieutenant-Colonel Morrice, having been wounded, as also Lieutenant-Colonel Elphinstone, 33rd Regiment, the command devolved upon Major Muttlebury, 69th Regiment, who marched the column round to the left on the outside of the place, leaving a wing of the 55th Regiment to collect the wounded on the glacis. This column then followed No. 1 column on to the Orange bastion.

Whilst General Cooke had been waiting at the Orange bastion the fate of the stormers of No. 4 column had been decided, and when the men of No. 2 column reached the rampart, he sent the 33rd Regiment to the assistance of the men at the Water-gate, but they came too late; part of the Regiment succeeded in joining the support of No. 4 column (the Royal Scots), which held its ground at the Water-gate, a portion inside and a portion outside the gate. The French, leaving a part of their force to hold the men at the Water-gate in check by the fire of the guns which they had turned on them, now attacked General Cooke's party on both flanks. Getting into a bastion adjacent to the Orange bastion, they turned the guns on the men of Nos. 1 and 2 columns, but were driven out by a bayonet charge of the wing of the 55th under Major Hog, and the 69th under Major Muttlebury. By this time day had broken. Guns having been turned on the men on the exposed rampart, General Cooke determined to withdraw, and a portion of the Guards managed to get out of the place; whilst they were leaving, the enemy again entered the bastion next the Orange bastion, and were again driven out by a charge of the 55th and 69th Regiments.

Finding he could not effect his retreat, General Cooke yielded to the summons of a French officer, who had come accompanied by Colonel Jones, 1st Guards, a prisoner and who told the General the fate of No. 4 column. **1814. Expedition to Bergen-op-Zoom.**

At daylight the Royal Scots, who had been exposed to a galling fire all night, and now stood with the swollen Zoom behind them, laid down their arms, and, together with part of the 33rd Regiment, were made prisoners.

Just as the troops had all surrendered the supports of the 2nd Division arrived outside the place, too late.

The losses in killed and wounded were as follows :—

| | Killed. | Wounded. |
|---|---|---|
| Officers .. .. .. .. | 17 | 54 |
| Sergeants .. .. .. .. | 29 | 28 |
| Drummers .. .. .. .. | 12 | 7 |
| Rank and File .. .. .. | 329 | 444 |
| Total .. .. | 381 | 533 |

Taken prisoners: Officers, 93; serjeants, 93; drummers, rank and file, 1,891. Total of all ranks, 2,077, many of whom were wounded.

The cause of this disaster was the failure of the officers in command of the various columns to carry out the orders they had received.

The fourth column encountered at first scarcely any resistance, and would probably have been able to have formed a junction with No. 1 column before the enemy attacked it in any force; or even if attacked before such a junction, it would have been in a better position to resist the onslaught of the French as a collected body than as a set of straggling detachments, and further, would have been within easy support from No. 1 column, when that column did penetrate the place.

Had General Cooke marched to his left according to his orders, instead of only sending a strong patrol in that direction, there seems no doubt that No. 4 column, reinforced by No. 1, could have successfully repelled the French attack, and the arrival of No. 3 column must have made the assault a success.

The garrison was about 2,800 strong, a portion of which would have been occupied at first with Nos. 2 and 3 columns; their attacking force, therefore, would not have exceeded by much the strength of Nos. 1 and 4 columns united, and when No. 3 arrived they would have been considerably outnumbered. Even at the eleventh hour, had the battalion of the Royal Scots held out, it is probable that on the arrival of the troops of the 2nd Division the place might still have been carried.

In a letter from Sir Thomas Graham to Lord Bathurst, dated Calmhout, 11th March, 1814, he says: "In short, the attack must

1814. Expedition to Bergen-op-Zoom.

have succeeded had the orders been obeyed. . . . We had considerable reinforcements at hand soon after daylight from the 2nd Division, when I had the mortification of seeing that they came too late. Still, had the Royals maintained the Water Post Gate, General Cooke would have held his ground, and the place must have fallen."

To whom the blame attaches for the late arrival of the men of the 2nd Division there is nothing to show.

On the 9th March the English prisoners were marched out of Bergen-op-Zoom to the Island of Tholen, and after a time were embarked for England, on condition that they should not serve again against France until an equal number of French prisoners in England had been exchanged. Three French officers and 119 men, made prisoners by the force under Sir T. Graham, were sent into Bergen-op-Zoom, a similar number of English prisoners being sent back in exchange.

After the repulse at Bergen-op-Zoom the troops retired into cantonments, investing Bergen, and preparing to invest Antwerp.

On the 21st March Fort Lillo was taken, with the loss of six killed and wounded.

Preparations were made to attack Fort Bath, in the island of South Beveland; but before they were completed, an armistice was agreed upon between Sir Thomas Graham and General Carnot, the Governor of Antwerp, the restoration of the Bourbons and the downfall of Napoleon having changed the aspect of affairs.

In accordance with the Convention of Paris of April 23rd, 1814, the city of Antwerp was occupied by British troops, the 2nd Division and the 1st Brigade of the 1st Division going into cantonments there, and garrisoning the various forts on the 5th May.

Shortly after this a portion of the British troops were marched to Brussels, and Sir Thomas Graham, now Lord Lynedoch, was placed in command of the various forces in Flanders, viz.: Prussian, Russian, Hanoverian, Swedish, Danish, Dutch, and Belgian, in addition to the British troops.

During the months of June and July, a force of Hanoverian troops in British pay arrived in Flanders, numbering some 3,000 bayonets and sabres. In the month of July, the Prince of Orange assumed command of the forces in Holland and Flanders, Lord Lynedoch returning to England.

The troops remained in various garrisons and cantonments in Holland and Flanders, being reinforced by British, King's German Legion, Hanoverian, and newly raised Dutch and Belgian corps, until the next year, when most of them took part in the Battle of Waterloo.

# INDEX.

www.ingramcontent.com/pod-product-compliance
Ingram Content Group UK Ltd.
Pitfield, Milton Keynes, MK11 3LW, UK
UKHW012239240726
13966UKWH00003B/1161

9 781845 741631